Wiring for Beginners

Ignite Your Creativity: Master The Art Of Wiring With Confidence And Joy. Learn How To Make Electrical Systems With Easy-To-Follow Explanations And Easy-To-Understand Images

By

MARVIN HALE

© **Copyright 2023 by MARVIN HALE All rights reserved.**

This document is geared towards providing exact and reliable information in regard to the topic and issue covered. The publication is sold with the idea that the publisher is not required to render accounting, officially permitted or otherwise qualified services. If advice is necessary, legal or professional, a practiced individual in the profession should be ordered.

From a Declaration of Principles, which was accepted and approved equally by a Committee of the American Bar Association and a Committee of Publishers and Associations. In no way is it legal to reproduce, duplicate, or transmit any part of this document in either electronic means or in printed format. Recording of this publication is strictly prohibited, and any storage of this document is not allowed unless with written permission from the publisher. All rights reserved.

The information provided herein is stated to be truthful and consistent in that any liability, in terms of inattention or otherwise, by any usage or abuse of any policies, processes, or directions contained within is the solitary and utter responsibility of the recipient reader. Under no circumstances will any legal responsibility or blame be held against the publisher for any reparation, damages, or monetary loss due to the information herein, either directly or indirectly. Respective authors own all copyrights not held by the publisher.

The information herein is offered for informational purposes solely and is universal as so. The presentation of the information is without a contract or any guarantee assurance. The trademarks that are used are without any consent, and the publication of the trademark is without permission or backing by the trademark owner. All trademarks and brands within this book are for clarifying purposes only and are owned by the owners themselves, not affiliated with this document.

Table of Contents

Introduction 6

Chapter 1: Tools and Equipment for Wiring 10
   1.1 Instruments absolutely necessary for wiring 14
   1.2 Different types of wires and cables 17
   1.3 Basic Rules and Guidelines for Wiring 21

Chapter 2: Building and Safety Codes for Electricity 25
   2.1 Overview of local electrical codes and regulations 27
   2.2 Compliance requirements for wiring installations 32
   2.3 Permits and inspections for wiring 35

Chapter 3: Residential Wiring Systems 40
   3.1 Service panels and circuit breakers for wiring 44
   3.2 Types of wiring systems (e.g., knob-and-tube, Romex) 49

Chapter 4: Basics of Wiring 54
   4.1 Wiring terminology and symbols 55
   4.2 Wire sizes and gauge selection 60
   4.3 Wiring color codes 63

Chapter 5: Wiring Techniques and Practices 66
   5.1 Proper wire stripping and termination 66
   5.2 Basic wiring connections (e.g., switches, outlets, lights) 69
   5.3 Grounding and bonding principles 72

Chapter 6: Wiring Projects 77
   6.1 Installing Switches and Outlets for Wiring 79

6.2 Wiring light fixtures and ceiling fans 85

6.3 Extending or adding new circuits 91

Chapter 7: Troubleshooting and Maintenance for Wiring 96

7.1 Identifying and resolving common wiring issues 98

7.2 Regular maintenance and safety checks 101

Chapter 8: Advanced Topics of Wiring 105

8.1 Wiring for specialized applications 105

8.2 Smart home wiring and automation 105

8.3 Energy-efficient wiring practices 109

8.4 Time-saving techniques for wiring projects 112

Chapter 9: ZigBee wireless technology and home automation 116

9.1 What kinds of capabilities do home automation systems provide their users? 117

9.2 What Is the ZigBee Wireless Technology, and How Does It Work? 119

9.3 ZigBee is Being Used for Home Automation 121

9.4 What exactly is meant by the term wireless home automation? 123

Chapter 10: Bonus Topics 129

10.1 Time-saving techniques for wiring projects 129

10.2 Avoiding common mistakes and pitfalls while wiring 132

10.3 Recommended resources for further learning 136

Conclusion 139

Introduction

Modern infrastructure is not complete without electrical wire, which provides the means for effectively and safely distributing power. It acts as the foundation of the power supply for residences, companies, and industries, allowing us to run our gadgets, light up our surrounds, and improve communication. Both professionals and individuals must have a basic understanding of electrical wiring in order to build, maintain, and troubleshoot electrical systems correctly. This book will examine the essential elements of electrical wiring, various wiring systems, safety precautions, and standard wiring techniques, all of which are necessary for the reliable and secure operation of electrical systems. We hope to give a thorough introduction to the wiring process and its importance in everyday life by exploring these subjects. An important consideration in determining whether or not the electrical system will function to the user's satisfaction and remain reliable over time is the caliber of the wiring. Wiring systems that are carelessly or incorrectly installed might be quite dangerous. These devices must be installed properly since a poor installation might result in fires and electrocutions.

You will get the chance to learn the crucial abilities required for efficient wiring operations with the aid of this course. An electrical circuit's many protective mechanisms, connections, and wires are all necessary circuit components. Anyone who works with electricity must make an effort to learn the fundamentals of these numerous components. A working electrical system is made up of a number of essential parts that make up electrical wiring. Conductors, insulation, connection boxes, toggles, and outlets are a few of these parts. Electric current may travel through materials called conductors, including copper and aluminum. They create the routes via which energy moves from the power generator to the final appliances. The conductors are encased in insulation, which shields them from unintentional contact and shocks caused by electricity by using substances like PVC or rubber. Junction boxes act as safe havens for electrical connections, reducing the possibility of electrical fires and making it easier to splice wires. While plugs, often referred to as receptacles, offer access to power sources for connected appliances and devices, switches are used to regulate the flow of energy to certain circuits or devices.

Different settings use various electrical system types, each with special qualities and functions. These include needle and tube wiring, conduit wiring, and non-metallic sheathed (NM), armored (AC), and NM cables. Individual conductors are supported by ceramic knobs and tubes in knobs and pipe wiring, which was historically typical in older dwellings. However, it is no longer a standard feature of new buildings and is often changed owing to safety concerns. Romex, another

name for NM cable, is widely utilized in residential settings. It is made up of ground wire and insulated conductors that are both covered in plastic. The usage of AC, or armored cable, in businesses and factories where toughness is essential provides improved mechanical protection. Individually insulated conductors are passed through either plastic or metal conduits during conduit wiring, which offers good resistance to heat, moisture, and mechanical stress. When dealing with electrical wiring, safety must always come first to avoid risks like fires and electrical shocks.

Maintaining human safety, as well as the dependability of the electrical system requires that proper safety practices be followed. Before attempting to repair any electrical circuit, these precautions include switching off electricity at the main electrical panel or fuse box. Avoiding unintentional contact with electrical wires that are alive is made easier by using a voltage meter to verify that the system is de-energized. Electrical systems must be properly grounded in order to protect against electrical issues and decrease the risk of electric shock. Regular checks and repairs are necessary to spot possible problems and fix them before they become risks. When establishing or changing electrical wiring systems, it is essential to follow accepted wiring principles. These procedures guarantee efficiency, safety, and adherence to laws and regulations governing electrical systems. Using the proper wire size for the present load is one of the standard wiring practices to avoid overheating and fire risks. Wires should be arranged and supported in a way that eliminates damage, lowers the chance of shorting out, and makes future maintenance easier. In order to prevent overloading, which may result in overheating and circuit failure, the electrical load should be distributed equally throughout the circuits. It's important to familiarize yourself with local electrical laws and regulations in order to ensure compliance with security standards and regulatory requirements. You are going to discover how to list the multiple techniques for connecting different sorts of wires and the numerous connections that are utilized for those lines by reading this book. Additionally, you will understand how to list the different links that are utilized for those lines. You will also learn how to list the numerous connections that are utilized for those wires in this lesson. You will have a grasp of the various thicknesses of cables and will be ready to list a variety of techniques for connecting wires and forming connections. In addition, you will have a concept of the many ways in which wires may be joined. You will be able to differentiate between one-phase and three-phase electricity, as well as grasp how transformers for one phase may be connected in either parallel or series configurations. In addition, you're going to be able to differentiate between one-phase and three-phase electricity. In addition to this, you will be able to differentiate between power that is delivered in single-phase and three-phase formats.

Chapter 1: Tools and Equipment for Wiring

When it comes to wiring, it is impossible to exaggerate how important it is to use the right tools and equipment. It doesn't matter whether you're working on a simple wiring job for your house or a complicated installation for a commercial building; having the appropriate equipment is very necessary in order to achieve safety, efficiency, and precision. In the following sentence, we will discuss the importance of the many tools and pieces of equipment that are needed for wiring.

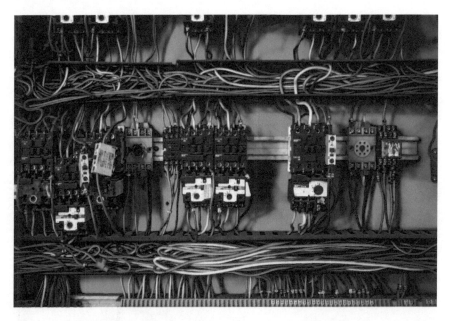

Using the appropriate equipment is the first and most important step in ensuring everyone's safety. Working in the electrical industry requires handling live wires and exposing oneself to potentially lethal currents. Insulated instruments, such as wrenches, pliers, and cable cutters, are meant to safeguard electricians from the dangers of receiving an electrical shock. The handles of these tools are insulated so that users do not come into direct touch with the conducting elements. This lowers the likelihood of injuries and collisions occurring. In addition, safety equipment like safety glasses, gloves, and insulated boots provide an additional layer of protection, protecting against possible accidents and electrical risks.

The use of the appropriate equipment is essential in order to do work that is accurate and exact, which is another essential component of electrical wiring. Tools for measuring, such as tape measurements, rulers, and levels, guarantee that measurements are accurate and that components are aligned correctly. When it comes to building conduits, mounting electricity boxes, or properly situating outlets and switches, this precision is of the utmost importance. By using the correct measurement equipment, mistakes such as misalignment outlets or incorrectly located junction boxes may be avoided, both of which have the potential to cause problems in the future. In addition, equipment such as wire strippers and wire cutters make it possible to strip and cut wires in a clean and accurate manner, which helps to ensure that suitable connections are made and reduces the likelihood of electrical malfunctions such as short circuits.

When wiring an electrical circuit, employing equipment that is specifically designed for the job is essential to maximizing productivity. The act of routing wires via conduits and other confined areas is made easier by the use of specialized tools, including fish tapes, duct benders, and wire-pulling grips. These technologies help the task go quicker and more efficiently, which saves both time and effort that might be better spent elsewhere. In addition, having the appropriate equipment readily accessible removes the need for creativity or the use of insufficient tools, both of which may result in errors, lost time, and a reduction in the overall quality of the job.

The general efficacy of a wiring job is also significantly impacted by a number of other factors, including the quality and longevity of the tools and equipment used. Making an investment in tools of a high grade assures their durability and dependability and reduces the frequency with which they must be replaced. Tools of high quality are manufactured to endure the rigors of wiring, such as the force required to screw in screws or the resistance encountered while cutting through thick insulation. Electricians are able to do their jobs with self-assurance and calmness when they use instruments that are dependable since they reduce the likelihood of tool failure through malfunction.

In addition, ensuring that one follows safety standards and laws is made possible by making use of the proper tools and equipment. The use of the appropriate equipment is essential to ensuring compliance with electrical regulations, which provide very stringent standards for wire installations. For example, torque screwdrivers guarantee that the electrical links are tightened to the needed standards. This prevents loose connections, which may lead to overheating or arcing, and ensures that electrical contacts are fixed to the required specifications. Not only is it essential to comply with rules for reasons of public safety, but also for reasons related to the law and insurance. The use of appropriate equipment and adherence to standard operating procedures in a sector are indicators of professionalism and a dedication to producing high-quality work.

In conclusion, having a toolbox or toolkit that is well-stocked with various tools displays both readiness and professionalism. It demonstrates to customers, managers, or regulators that the electrical contractor takes their profession seriously and places a high emphasis on both quality and safety. Electricians are provided with the ability to perform a wide variety of wiring chores in an expedient and effective manner by virtue of having access to a broad array of tools and equipment.

1.1 Instruments absolutely necessary for wiring

When beginning a job involving wiring, having access to the appropriate instruments is essential for ensuring productivity, precision, and personal safety. The following is an explanation of the functionality of some of the most important tools for wiring projects:

Screwdrivers: To remove or tighten the screws that are found in electrical outlets, toggles, and other devices, screwdrivers with shielded handles are used. In order to handle a wide variety of screw types, they are available in a wide range of sizes and kinds, including both flathead and Phillips heads.

Pliers: Pliers are multipurpose hand tools that may be purchased in a variety of configurations, including lineman's tools, cable strippers, and needle-nose pliers, amongst others. While lineman's pliers are employed for holding and cutting wires, needle-nose pliers are great for getting into tight locations and handling tiny components. Lineman's pliers are utilized for grabbing and cutting wires. Wire strippers are instruments that are manufactured for the sole purpose of removing the insulating material from electrical cables without causing any harm to the conductor.

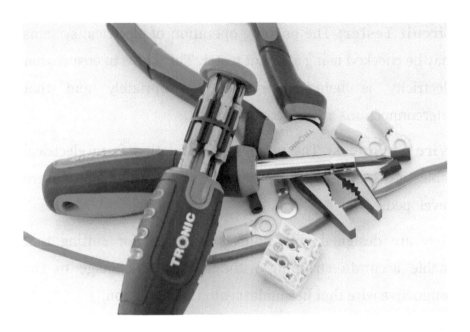

Wire Cutters: Wire cutters, also known as diagonal cutters, are tools that are used for the purpose of cutting electrical wires in a manner that is both clean and effective. They have shrewd cutting edges that let wires be cut precisely to the correct length, and they come in a variety of sizes.

Voltage Tester: When determining whether or not a circuit is active or de-energized, a voltage tester or multi-meter is often utilized. Before repairing electrical connections, it helps ensure that the appropriate safety procedures are performed, which helps avoid inadvertent shocks.

Fish Tape: Fish tape is a lengthy and flexible tape that is used for the purpose of routing wires through pipes, walls, or ceilings. It is helpful for pulling cables in regions that are difficult to reach or occluded.

Circuit Tester: The ongoing operation of electrical systems may be checked using a circuit tester. This helps to ensure that electricity is being distributed appropriately and that interconnections are not compromised.

Wire Strippers: The removal of insulation from electrical cables is the primary purpose for which wire strippers were developed.

They are designed with notches or edges for cutting that enable accurate stripping without causing damage to the conductive wire that lies underneath the insulation.

Electrical Tape: Insulating electrical connections and keeping them secure requires the use of a product that is indispensable: electrical tape. It offers an additional layer of defense against unforeseen touch, abrasion, and wetness.

Fish Rods: Fish rods, often referred to as wire rods for fishing, are flexible rods that are used to direct and route cables through walls, pipes, and other tight locations. Fish rods are additionally referred to as wire fishing rods. They assist in navigating wire routes and reaching regions that are inaccessible.

Circuit Breaker Finder: An instrument that assists in locating certain circuit breakers inside an electrical panel is referred to as a "circuit breaker finder." When working on certain circuits, the process of selecting the appropriate breaker to turn off is streamlined as a result of this feature.

Cable Stapler: A cable staple is a tool that is used for the purpose of securing electrical wires and cables to surfaces such as ceilings or studs in order to guarantee appropriate organization and avoid possible risks.

Non-Contact Voltage Detector: A portable device known as a non-contact energy detector is one that can detect the existence of electricity without making physical contact with the source of the voltage. During electrical work, it assists in locating live wires or circuits that are not functioning properly, hence increasing worker safety.

These tools are only a sampling of the many different kinds of equipment that are available for use in wiring tasks. It is possible that extra tools and equipment will be required in order to complete the current work successfully. When dealing with electrical systems, you should always make sure that you have the necessary equipment and that you follow all of the safety standards.

1.2 Different types of wires and cables

For the purpose of electrical wiring, there is a wide variety of cables and wires available, each of which is tailored to satisfy a certain set of criteria and functions. The following is a list of a few of the most frequent varieties of cables and wires that are used in electrical wiring projects:

- **Non-Metallic Sheathed Cable (NM or Romex)**

NM cable is made up of at least two insulated wires, which are commonly black and white in color, as well as a bare copper grounding wire, and all of these components are contained in a polymer sheath. It is well-suited for usage in dry regions and is often put to use in home wiring applications.

- **Armored Cable (AC)**

AC cable, which may also be referred to as BX or elastic metal conduit, is comprised of separate insulated wires that are encased inside an elastic metal sheath. The metal sheath, which is utilized in a variety of commercial and industrial applications, offers safeguards against physical damage and is often used in these contexts.

- **Underground Feeder Cable (UF)**

UF wire was developed specifically for use in subterranean and outdoor electrical installations. It can be buried straight in the ground eliminating the requirement for conduit since it is resistant to moisture and comprises insulated wires that are covered in a sheath that is also resistant to moisture.

- **Coaxial Cable**

The transmission of high-frequency signals, most often those used for broadcast and internet connections, is accomplished via the use of a coaxial cable. It is made up of an insulating core, a metal barrier, and an outside insulating layer. The conductor is located in the middle of the insulating core.

- **Ethernet Cable**

Networking and the transport of data are two of the primary applications for Ethernet cables, which are additionally referred to as Cat5e, Cat6, and Cat7 connections. They have numerous pairs of wires made of copper twisted together, which allows for the flow of data at a fast pace in a reliable manner.

- **Single-Conductor Wires**

Wires with a single conductor are made up of a single strand of copper, which is often encased in insulation. Applications like grounding, appliance wiring, and branching circuit wiring are frequent places to find them in use.

- **Speaker Wire**

Speakers can only be connected to audio systems via the use of speaker wire. Typically, it is made up of two conductors with strands that are color-coded for convenience in determining their identity.

- **Thermostat Wire**

A thermostat wire is a wire that has been developed especially for the purpose of connecting HVAC (heating, ventilation, and air conditioning) systems. In most cases, it is composed of a number of conductors, each of which serves a distinct function, such as regulating the temperature, providing power to the thermostat, or sending signals.

- **Fiber Optic Cable**

The transmission of data across great distances and at high speeds is made possible by using fiber optic connections. They are made up of very fine strands of either glass or plastic fibers, and instead of carrying electrical currents, they transmit light messages.

- **High-Temperature Wire**

Wires rated for use at high temperatures are constructed to endure the intense heat that may be found in some situations, such as industrial applications, ovens, and furnaces. They have walls lined with substances that can resist high temperatures without breaking down.

Each kind of cable or wire has a unique set of properties that confer upon it a particular set of advantages when used in a certain context. It is essential to choose the appropriate kind of cable or wire based on a variety of criteria, including the environment, the voltage needs, the current capacity, and the particular electrical standards and laws in your region.

1.3 Basic Rules and Guidelines for Wiring

When it comes to installing wire, there are a few basic rules and principles that ought to be followed to guarantee security, effectiveness, and conformity to electrical standards. These rules and guidelines may be broken down into numerous categories. The following are some fundamental guidelines for wiring:

- **Safety First**

When doing any kind of job involving electricity, putting safety first is the single most important guideline. Before doing any electrical work in a circuit or other location, you should always cut off the electricity supply to that circuit or area. Make sure that you are using the correct personal protective equipment (PPE), such as gloves with insulation and safety eyewear. Make sure that the instruments and equipment they use are in adequate condition and appropriate for the job you are doing.

- **Compliance with Electrical Codes**

Electrical codes are rules that are enacted by authorities to guarantee that electrical installations are carried out in a safe manner. While working on the wiring, you should familiarize yourself with the applicable codes and stick to them. The size of the wires, the positioning of the outlets, the grounding, and any other relevant safety factors may be outlined in the codes.

- **Plan and Design**

It is important to properly organize and create the layout before beginning any job involving wiring. Think about the necessary electrical components, the calculations of the loads, and the placement of the outlets, switches, and light fixtures. To assist you in the installation process, draw out a diagram of the wiring or schematic.

- **Use the Proper Wire Size**

When it comes to safety and avoiding potential electrical mishaps, choosing the appropriate wire size is quite necessary. The amount of current (amperage) that a circuit needs will determine the size of the wire that is needed. Determine the suitable wire gauge for the particular application by consulting either the NEC (National Electrical Code) or the rules in effect in your area.

- **Avoid Overloading Circuits**

Every circuit has a certain maximum load that it can securely carry at one time. In order to avoid overheating, damaging the circuit, and increasing the risk of a fire, you must not go over this capacity. Determine the overall amount of electrical load that is being carried by a circuit and, if necessary, assign it to the appropriate number of other circuits.

- **Proper Grounding**

For reasons of both safety and protection against shocks from electricity and damage to equipment, grounding is a vital practice. Make sure that the grounding of all the electrical circuits and gadgets is done correctly. Put in grounding cables, grounding rods, and make sure all of the electrical outlets are connected to the grounding network.

- **Adequate Conduit and Wiring Protection**

Use adequate conduits, trays for cables, or raceways to safeguard wire from damage caused by direct contact with other objects. These shield the wires from accidental strikes and the effects of the surrounding environment. In addition, it is important to keep electrical lines away from electrical sources of water, heat, and any other potentially dangerous circumstances.

- **Proper Wire Connections**

Establish safe and dependable connections between the wires to guarantee enough electrical conductivity and minimize the possibility of the connections coming loose. After using wire strippers to remove the outer layer of cable insulation to the appropriate length, firmly twisting the wires together, and using wire connections or terminal protectors to seal the connection, you are finished.

- **Labeling and Documentation**

Label and keep a record of all wire connections, connection boxes, and wiring structures in the appropriate manner. Future servicing, troubleshooting, and adjustments are made easier as a result of this. Circuits, toggles, outlets, and other elements may be easily identified with the use of color-coded cables and labels.

- **Regular Inspections and Maintenance**

After the installation of the wire has been finished, you should conduct routine inspections to look for any indications of wear, destruction, or possible risks. Check for indicators of overheating, such as loose connections, broken wires, and melted insulation. Maintaining a secure and dependable electrical system requires prompt resolution of any problems that may arise. It is essential to keep in mind that wiring should almost always be carried out by certified electricians or other personnel who have the appropriate level of education, training, and experience.

Chapter 2: Building and Safety Codes for Electricity

There are several justifications for why electrical rules and regulations are absolutely necessary. First and foremost, they place a high value on the protection of both people and property. Electricity is not without its dangers; among them are the possibility of receiving an electric shock and starting fires. Electrical codes are sets of principles and standards that are established for the planning, setting up and upkeep of electrical systems. The purpose of these codes is to reduce the risks associated with using electrical systems and to guarantee that using them is safe. Maintaining compliance with electrical regulations may assist in the prevention of accidents, injuries, and damage to property. The second benefit of electrical rules is that they encourage uniformity and consistency in the installation of electrical equipment. They provide a uniform structure that guarantees processes are consistent across a variety of projects, regions, and governments. Electricians, suppliers, and engineers may guarantee that power lines are built and operated in a reliable and consistent way by complying with these rules. This is true regardless of the unique conditions that are present.

The ability to guarantee interoperability and compatibility is yet another significant advantage of having electrical codes. The standards for electrical equipment, gadgets, and components are defined by the electrical codes, which ensures that these things will work together without any problems. These codes define the standards for electrical wiring, connections, outlets, toggles, and other devices, making it possible for different electrical components to be compatible with one another and safely connected to one another. In addition, technical advances and shifting business practices are taken into consideration while developing electrical regulations. The National Electrical Code is revised periodically to take into account any new developments in terms of technology, materials, or processes. This guarantees that electrical installations remain up with innovations in the industry, adopting the most recent best practices, security protocols, and technology that are efficient in their use of energy. The adoption of innovative ideas while still putting an emphasis on safety and dependability is made possible, in large part, by the existence of codes.

Electrical codes and standards, in addition to promoting safety and uniformity, also assist in maintaining compliance with various laws and regulations. Generally speaking, adherence to these rules is a legal obligation that is enforced by the respective local governments and regulatory agencies. Individuals and businesses are able to avoid the penalties, monetary penalties, and legal obligations that are connected with non-compliance with electrical rules when they adhere to such standards. In addition to this, it serves as a foundation for the inspections, permits, and certificates that are essential to ensure that electrical installations are in compliance with the law and satisfy the necessary standards. In addition, experts in the business have access to advice and reference materials via the use of electrical codes. These codes are relied on by electricians, architects, designers, and other professionals engaged in electrical work to guarantee that the designs, structures, and maintenance operations they do are in accordance with the industry standards that are currently in place. Codes are very useful resources since they provide direction on the appropriate procedures, components, and methods to adhere to while working on electrical projects.

2.1 Overview of local electrical codes and regulations

Building codes and electrical safety rules are important sets of laws and standards that control the installation, utilize, and upkeep of electrical wiring in buildings. These regulations and guidelines ensure that electrical systems are safe to install, use, and maintain. The inhabitants' safety, the prevention of electrical risks, and the promotion of electrical systems that are efficient and dependable are the primary focuses of these regulations. In order to safeguard individuals, assets, and the general health and prosperity of a community, adherence to these regulations is very necessary. The National Electric Code (NEC), which is produced by the National Fire Prevention

Association (NFPA), is considered to be one of the building codes that has the highest level of public acceptance in the United States. The NEC is the major electrical code used by numerous states and local authorities since it establishes the minimum requirements for safe electrical activities and is approved by those jurisdictions. The National Electrical Code (NEC) is continuously revised to take into account newly developed electrical technologies and solve any newly discovered dangers.

The National Electrical Code applies to a diverse array of electrical systems, including those found in homes, businesses, and industrial facilities. It contains rules for electrical connections, grounding, overcurrent security, electrical panels, outlets, lights, and other electrical components. Other topics covered include illumination. The National Electrical Code (NEC) must be followed in order to guarantee that electrical systems are correctly planned, installed, and maintained. At the national, regional, and municipal levels, numerous organizations and institutions, in addition to the NEC, are responsible for establishing construction and safety rules pertaining to electrical systems. These codes may include certain standards and rules that are based on regional characteristics, climatic conditions, and the sorts of buildings that are there. Before beginning any kind of electrical work, it is essential to check into the appropriate norms and regulations that are applicable to the particular jurisdiction in question. The following is a condensed version of the major goals that construction and safety rules for electricity strive to achieve:

- **Electrical System Safety**

The protection of both persons and property from potential electrical dangers is the primary purpose of these regulations. They define rules for correct grounding, insulating material, and overcurrent protection in order to avoid electric shocks, fires, and other possible risks. These recommendations are intended to safeguard against potential hazards. To further improve the security of electrical systems, codes demand the installation of tamper-resistant locations, arc fault network interrupters (AFCIs), and ground-failure circuit interrupters (GFCIs), among other safety precautions.

- **Proper Installation Practices**

Electrical systems may only be installed properly if they comply with the relevant codes, which provide exhaustive instructions and prerequisites. They provide the necessary guidelines for the methods of wiring, the sizes of the wires, the use of conduit, and the connecting procedures. The proper execution of electrical installations is ensured by following these recommendations, which also serve to reduce the likelihood of defects, failures, and harmful circumstances occurring.

- **Load Calculations and Capacity**

The criteria for electrical load estimates and capacity that are set out by the building codes apply to a wide variety of structures and locations. These estimates take into account a variety of parameters, including the estimated power usage as well as the number of locations, lighting fixtures, and appliances. Overloading, voltage dips, and failures of electrical systems are avoided by following the rules because they ensure that the electrical infrastructure is capable of safely handling the intended load.

- **Energy Efficiency**

Provisions that encourage energy conservation in electrical wiring are increasingly becoming standard in many building codes. This includes regulations for lights, controls, and equipment that are efficient in their use of energy. These steps are designed to decrease utility expenses, cut down on energy usage, and lessen the negative influence that electrical installations have on the surrounding environment.

- **Accessibility and Usability**

The accessibility and usefulness of electrical systems are addressed in building regulations, with particular attention paid to those who have physical limitations. They provide criteria for the location and height of power outlets, toggles, and other controls in order to guarantee that all individuals are able to readily reach and use them.

- **Inspections and Permits**

Electrical installations are often subject to inspections and permit requirements imposed by building regulations. This guarantees that installations are examined by trained specialists and comply with the criteria of the applicable codes. Before the electrical infrastructure is placed into operation, inspections must be performed to ensure that the work has been completed to the required level of safety and to offer a chance to correct any defects discovered.

- **Ongoing Maintenance**

The need to perform routine maintenance and inspections on electrical systems is emphasized by the codes to guarantee that these systems continue to function in a safe and dependable manner. They may stipulate regular inspections, tests, and maintenance to be performed on all electrical components and equipment. Building codes and electrical safety regulations are very important in protecting people's lives, encouraging the responsible use of energy, and ensuring that the reliability of electrical wiring in buildings is preserved. It is vital for building owners, architects, engineers, and electricians to comply with these norms in order to guarantee that their electrical systems are safe.

2.2 Compliance requirements for wiring installations

Compliance requirements for wire installations are the rules, guidelines, and norms that must be adhered to while designing, constructing, and upkeep electrical wiring systems. These criteria must be met before an installation can be considered compliant. These regulations have been put in place to safeguard the safety of electrical systems while also maximizing their efficiency and dependability. There are a few essential components that are often anticipated for wire installations, despite the fact that particular compliance standards might differ based on the area of jurisdiction and the kind of installation that is being performed. Here are some typical compliance criteria to consider:

- **National and Local Electrical Codes**

It is very necessary to ensure that both national and local electricity codes are adhered to. These regulations, such as the NEC (National Electrical Code) in the United States, provide an overview of the standards that must be met by electrical installations at the very least. They cover a variety of topics, such as the appropriate size of wire to use, grounding, and overcurrent protection, installing electrical panels, and choosing the appropriate equipment. By adhering to these guidelines, one may guarantee that the installation complies with the legal requirements and is up to the recognized safety standards.

- **Wire Sizing and Ampacity**

In order to maintain compliance, it is necessary to choose the correct wire size for the circuit by basing your decision on the required amount of current. When determining the appropriate size of a wire, it is important to take into account the projected load, the temperature in the environment, and the total length of the cable run. To achieve compliance, one must follow the ampacity tables and other requirements given in the electrical code. This is done to guarantee that the cable can safely carry the intended current without being overheated.

- **Wiring Methods and Materials**

The various kinds of installations each have their own set of compliance standards, which detail the allowed wiring techniques and materials. These criteria specify the kinds of cables, conductors, and conduits, as well as the insulating materials, that are allowed for use in certain applications. For the purpose of ensuring the reliability and security of the installation, the techniques of wiring must be selected and implemented in line with the relevant regulations of the code.

- **Grounding and Bonding**

Grounding and bonding procedures must be carried out correctly for compliance. The creation of an electrical fault channel with low resistance is the goal of grounding systems. This helps to ensure that excessive electrical energy is routed away from people and securely into the ground. In order to ensure compliance, grounding and bonding criteria must be met for electrical panels, devices, and conductive materials. These requirements are designed to reduce the likelihood of electrical failures and electrical shocks while also protecting equipment from damage.

- **Overcurrent Protection**

In order to be compliant, an adequate overcurrent protection device, such as a fuse or circuit breaker, must be installed in the electrical system. These devices prevent dangerously high levels of electricity from flowing through the wire and the electrical equipment. The appropriate rating and kind of overcurrent safeguards are specified by the compliance criteria. These requirements are deduced from the wire size, the load, and the particular application.

- **Proper Installation Techniques**

To achieve compliance, one must follow the appropriate installation procedures in order to guarantee that the wire is adequately supported, securely attached, and shielded from any potential physical harm. This involves following requirements for bending and fastening wires, spacing of outlets, toggles, and junction containers, and, where applicable, adhering to fire-resistant building methods. In order to guarantee dependable and secure electrical connections, compliance necessitates the use of correct techniques of termination and connection.

- **Safety Devices**

Installation of safety equipment per code specifications is considered compliance. Some examples are GFCI outlets, which prevent shocks from electricity in wet or damp areas, AFCI outlets, which prevent fires from arcing faults in electrical wiring; and tamper-resistant outlets, which prevent foreign objects from being accidentally inserted into outlets. Compliance guarantees that these safety measures are set up properly and according to standards.

- **Documentation and Labeling**

Wiring installations must be documented and labeled often to ensure compliance. Electrical panels, wiring, and components must be properly labeled for simple identification and maintenance. Electrical plans, as-built drawings, inspection summaries, and upkeep logs may also be part of the documentation package.

2.3 Permits and inspections for wiring

In the process of installing wiring, obtaining permits and passing inspections are essential steps that must be taken.

They play an important part in ensuring that the electrical work is being carried out in a safe manner, that it complies with the relevant norms and regulations, and that it fulfills the required standards. The acquisition of licenses and participation in inspections are important actions that contribute to the protection of persons, their possessions, and the community as a whole. Let's get into the nitty-gritty of obtaining licenses and passing inspections for installing electrical wiring. Permissions, often known as permits, are legal authorizations that may be given by the necessary regulatory agencies or by municipal construction departments. They provide the function of formal authorization to carry out the designated electrical work. Permits are required in order to fulfill the responsibility of ensuring that a planned wire installation is in accordance with all of the relevant rules, standards, and security requirements. In order to acquire a permit, a person is often required to provide extensive paperwork on the proposed project. This documentation may include electrical drawings, load calculations, gear specifications, and any other pertinent data.

In most cases, the procedure of applying for a permit entails having the planned wire installation evaluated by a representative from the building authority or another regulatory body. With this assessment, we can confirm that the project complies with all of the essential standards and norms. After the application has been reviewed and accepted, the permit will be granted, at which point the electrical work will be given the green light to continue. It is essential to be aware that engaging in electrical work lacking the proper authorization may result in monetary penalties, administrative fines, and even possible legal repercussions. Inspectors who are competent to do so are often hired by the local construction authority or regulatory body to carry out the inspections. Inspections are carried out with the goal of ensuring that the wire installation was carried out appropriately, that it conforms to the designs that were granted approval, and that it complies with all of the relevant laws and regulations. Inspections are often carried out at different phases of the construction endeavor, such as before hiding the wiring after the rough-in work has been completed and once the project has been brought to its conclusion. The particular inspections that must be performed on wire installations are likely to differ from one jurisdiction to another and from one kind of project to another. However, the following are some things that are normally checked during routine inspections:

1. **Rough-In Inspection**

This examination takes place after the basic wiring has been finished. This includes the installation of electrical boxes, the running of cables or conduits, and the placing of switches and outlets. The goal is to make sure that the cabling is done correctly, that it complies with the criteria of the code, and that it is prepared for additional work.

2. Service Panel Inspection

A separate check will be performed on the service panel, which is part of the electrical system that contains the main power disconnect as well as circuit breakers or fuses. During this examination, it will be determined whether or not the panel was appropriately placed, whether or not it is properly anchored, and whether or not it satisfies the essential safety requirements.

3. Final Inspection

After the electrical installation has been finished in its entirety, the final inspection will be performed. It entails examining the electrical system as its whole to determine whether or not it complies with all applicable codes, whether or not the grounding and bonding are done correctly, whether or not the wiring techniques are appropriate, whether or not there is sufficient protection against overcurrent, and whether or not the switches, power sources, and other components are operational. During assessments, the inspector looks at the electrical system from a variety of angles and perspectives. They analyze the location of devices and elements, assess the quality of the workmanship, verify conformity with authorized designs, check for correct wire size and dismissal, examine bonding and grounding, assess the quality of the grounding, and certify the installation of security devices such as GFCIs and AFCIs when applicable.

Before final clearance may be given, any flaws or code violations that are found during inspections need to be fixed in order for the project to be considered complete. In most cases, the inspector will submit a written report that details any adjustments or repairs that are required. When all of the necessary alterations have been implemented, the next phase of the process, which involves confirming compliance and issuing the last authorization or certificate of occupancy, may be arranged. It is essential to understand that acquiring licenses and passing inspections are not only mechanical steps in a bureaucratic process. They are devised with the purpose of protecting people, properties, and the neighborhood as a whole. By acquiring permits and being inspected, electrical work is put under close scrutiny by industry professionals to verify that it complies with all of the necessary regulations and is up to the required level of safety. This contributes to the prevention of electrical risks, the identification of possible problems, and the promotion of the general integrity and dependability of the wire system.

Chapter 3: Residential Wiring Systems

Within residential structures like homes, apartments, and condos, residential wiring structures refer to all the electrical wiring and equipment that provide electricity and allow electrical appliances and systems to work properly. These residential wire systems may also be referred to as domestic electrical systems. These systems consist of a network of wires, devices, and other components that distribute electricity throughout the home, beginning at the primary electrical service entry and moving to numerous rooms, spaces, and appliances as it travels through the structure. The electrical requirements of homeowners and tenants are taken into consideration when residential electrical wiring systems are developed. This allows for the electricity to be distributed throughout the property in a manner that is both safe and effective. They ensure that the lights, appliances, cooling and heating systems, multimedia devices, and various other electrical equipment may be powered and operated inside the domestic environment by serving as the pillars of the electrical connection system. The electrical utility, the main entrance, electrical sheets, circuits, wires, outlets, switches, and numerous gadgets that promote safety and control are the essential elements that makeup home wiring systems. Let's take a closer look at each of these elements one at a time:

- **Electrical Service Entrance**

The point at which the electrical power supplied by the utility company is brought into the home is referred to as the electrical service entry. A meter, a main separation and service entry wires are normally included in its components. A secure and dependable connection to the power grid is ensured by the electrical service entry.

- **Electrical Panels**

Electrical panels, which also go by the names distribution panels and breaker panels, are the components of a home that are in charge of delivering electrical power to the various rooms. They take the power that is provided by the utility company at the service door and then distribute it to the various circuits in the building using fuses or circuit breakers.

Controlling and safeguarding the electrical system may be done from a single place thanks to the presence of electrical panels.

- **Circuits**

Circuits are the paths that electrical current travels down in order to go from the power supply to the numerous outlets, toggles, and appliances located throughout the home. Residential wiring systems often consist of many circuits, each of which serves a distinct set of electrical loads and spaces. Lighting wiring, general-purpose wiring, specific circuits for high-power gadgets, and customized circuits for particular uses like HVAC systems and home theaters are all examples of common types of circuits.

- **Wiring**

The wiring in a home is the network of wires that transports electric current from the power supply to the various outlets, switches, and other devices located throughout the home. A wide variety of cables, including non-metallic wrapped cable (NM cable) and armored cables (AC), as well as single conductors contained inside conduits, are included in this category. The dimensions of the wire are determined by the electrical load demands as well as the safety criteria stipulated in the relevant electrical codes.

- **Outlets and Receptacles**

Electrical power may be accessed at various points around the home through the outlets and receptacles that have been installed there. They make it possible to connect various electrical products and gadgets to a power source. Standard wall locations, ground-fault wire interrupter (GFCI) connectors in regions where water is around, and arc-fault wire interrupter (AFCI) receptacles to prevent fires that are caused by spiraling faults are all types of outlets that may be installed in a building. Standard wall outlets can also be installed.

- **Switches**

Controlling the flow of electricity to various devices, such as ceiling fans, lighting fixtures, and other appliances, is made possible by switches. They provide residents the ability to switch lights off or on, change the speed of fans, and regulate a variety of other electrical operations.

Different applications call for different kinds of switches, such as single-pole, three-way, or dimming switches. This is determined by the functionality that is needed.

- **Safety Devices**

In order to safeguard residents from potential electrical dangers, residential wiring systems consist of a variety of different safety measures. These safety measures include grounded circuit interrupters (GFCIs) that monitor the flow of electrical current, identify any imbalances, and immediately cut power to cut off the flow of electricity and avoid electric shocks. Arc fault circuit interrupters, often known as AFCIs, are devices that are intended to identify potentially hazardous arcing faults and interrupt power supplies in order to avoid electrical fires. It is essential that residential wiring systems adhere to all relevant electrical rules and safety requirements to guarantee both the installation's integrity and its ability to perform its intended functions. In order to be in compliance with the rules, you need to utilize permitted wiring techniques and materials, use wires of the correct size, properly ground and bind them, and guard against excessive current. It is crucial to have certified electricians who are familiar with the particular needs and safety procedures connected with home electrical installations and install and maintain residential wiring systems. This is because the particular demands and safety standards involved with residential electricity installations may vary greatly.

3.1 Service panels and circuit breakers for wiring

In residential as well as business, electrical wiring systems, service panels, and breakers are necessary components that cannot be left out. They are essential components in the system that ensures the secure transmission and protection of electricity inside a structure. Service panels, which are additionally known as electricity panels or breaker panels, serve as the primary hub for the distribution of electrical power into a building, and circuit breakers are there to prevent overcurrent and electrical problems. Service panels may also be referred to as breaker panels. Let's take a closer look at each of these elements in more detail:

- **Service Panels**

Bus bars, neutral and grounding bars, circuit breakers, and fuses are some of the electrical components that may be found inside a service panel, which is a metal box containing these components. The electrical power that comes into a building from the power company is routed via this component, which acts as the building's central distribution point. The following are some of the primary uses for a service panel:

- **Power Distribution**

The service panel is responsible for distributing the electrical power that is brought into the building from the energy provider's entry to the numerous circuits that are located throughout the structure. It does this by separating the electrical demand into its component circuits and ensuring every circuit gets the proper amount of power in accordance with the function that it was designed to serve.

- **Electrical Service Entrance**

The service panel is the location where the connections for the electrical service entry wires are made. These conductors, which are commonly known as service entry cables or company conductors, are responsible for bringing electrical energy from the distribution network of the utility company into the building. These conductors may connect to the service panel in a manner that is both safe and secure, thanks to the panel's design.

- **Main Disconnect**

The building's electrical power may be completely cut off via the use of a main disconnect button or breaker, both of which are located inside the service panel. It is important to have access to this main disconnect in the event of any electrical faults or issues, as well as in the case of any emergency scenarios.

- **Bus Bars**

Within the service panel, the bus bars are the metal bars that act as the primary conductors for transmitting electrical energy to the circuit interrupters and fuses. Bus bars are also known as busway bars. They are responsible for distributing the current that is carried from the conductors of the incoming service to the many individual circuits located throughout the structure. The service panels' enclosure is normally constructed out of copper or aluminum, and it contains bus bars that are installed in a safe and secure manner.

- **Neutral and Grounding Bars**

The service board also includes grounding and neutral bars in its construction. While the earthing bar offers an interface for the conducting conductors, the neutral bar acts as a common connecting spot for the neutral cables coming from the various circuits. These bars guarantee that the grounding is done correctly and contribute to the continued stability and safety of the electrical system.

- **Circuit Breakers**

In the case of an overcurrent, an electrical short circuit, or other electrical defects, circuit breakers are protection devices that are meant to instantly separate electrical circuits. They provide an essential function for the purpose of ensuring the protection of electrical wire systems. The following is a list of the most important characteristics and uses of circuit breakers:

- **Overcurrent Protection**

Circuit breakers prevent excessive current flow in electrical circuits, which may otherwise cause overheating, which in turn could result in damaged equipment or even fire dangers. They are constructed in such a way that they will either trip or release the circuit in the event that the current surpasses the breaker's maximum rating. By doing so, the destruction of the electrical wiring may be avoided, and the likelihood of experiencing electrical risks is reduced.

- **Tripping Mechanism**

When an overcurrent situation is recognized, the circuit is broken off using a variety of different tripping mechanisms that are built into circuit breakers. The three kinds of circuit breakers that are most often used are thermal, magnetized, and combined thermal-magnetic. Breaks classified as thermal depend on the heat produced by the current, breaks classified as magnetic react to the magnetic attraction caused by a fault, and breakers classified as thermal-magnetic combine the two principles.

- **Ampere Rating**

There is a wide range of ampere ratings for circuit breakers, each of which indicates the maximum amount of current that the device can safely manage. It is essential to carefully pick circuit breakers that have sufficient ampere ratings in order to guarantee that the electrical systems are appropriately protected and capable of carrying the anticipated loads in a secure manner.

- **Circuit Identification**

It is common practice to label or number circuit breakers so that it is clear which circuits or parts of the structure they are responsible for protecting. In the event that there are electrical problems or maintenance work to be done, this enables simple identification and troubleshooting. Electrical contractors or building occupants may rapidly discover and isolate individual circuits with the assistance of clear labeling.

- **Arc Fault and Ground Fault Protection**

Circuit breakers with advanced technology may be equipped with supplementary safety features like arc fault circuit interrupters and ground-failure circuit interrupters, respectively. Electrical fires may be avoided by installing AFCIs since they can identify potentially hazardous arcing faults and rapidly disconnect the circuit. Protecting against grounding faults, which may happen when current passes from a circuit of electricity to a grounded appear, such as an individual or a moist place, is the primary function of ground fault circuit interrupters, or GFCIs. Electrical wiring systems are not complete without the presence of important components such as service switches and circuit breakers. Circuit breakers are used to prevent current overload and electrical problems, while service panels are the points in a building that are responsible for the distribution of electrical power to the rest of the building. Together, they guarantee that electrical power is distributed across residential and commercial buildings in a manner that is both safe and as effective as possible.

3.2 Types of wiring systems (e.g., knob-and-tube, Romex)

There are many different kinds of wire systems that are used in homes and businesses, and each of these systems has its own set of features, materials, and methods of installation.

The following is a concise description of several popular kinds of electrical wiring systems:

1. Knob-and-Tube Wiring

Knob-and-tube cabling is an earlier kind of electrical wiring that can often be found in buildings that were built before the 1930s. It is made up of separate conductors made of insulated copper that are held together by ceramic knobs and tubes. The conductors are routed individually, with sufficient distance between the hot and ground wires, which are held in place by ceramic knobs. This style of wiring needs careful installation in order to preserve the appropriate spacing and avoid overheating. Knob-and-tube electrical frequently gets replaced with more contemporary wiring systems because of its advanced age and the possible safety risks it poses.

2. Romex Wiring (Non-Metallic Sheathed Cable)

It is usual practice for people to refer to the non-metal wrapped cable (also known as NM cable) by its brand name, Romex. It is one of the kinds of wiring systems that are utilized in contemporary residential and commercial structures and is one of the most common. Romex is made up of two or more shielded conductors, which are often made out of copper. These conductors are then encased in a non-metallic sheath, which is typically made out of PVC or polymer. It comes in a variety of shapes and sizes, including 2-wire (with both warm and neutral wires) or 3-wire (as well as an extra ground conductor), all of which have the option to be purchased. Romex is popular for usage in a variety of general-purpose wire applications because it is simple to manipulate.

3. BX Cable

A form of wire system known as BX cable, which is additionally referred to as cable with armor (AC), is characterized by the presence of a flexible metal sheath that surrounds insulated conductors. The metallic sheath not only offers the necessary protection against mechanical damage but also acts as grounding conductivity. In businesses and factories that demand additional durability and safeguarding against damage, BX cable is often utilized because of its versatility and reliability.

4. Conduit Wiring

The process of wiring a conduit includes threading individual conductors that are insulated through tubes made of metal or plastic. The wires are protected and contained inside conduits, which are hollow tubes built for this purpose. This form of wiring provides increased protection against the effects of physical damage, as well as moisture and corrosion. Conduit wire is a kind of enclosed wiring that is often used in commercial and industrial settings, as well as in locations where the electrical system is either exposed or subjected to severe weather.

5. Aluminum Wiring

During the period between the 1960s and the 1970s, aluminum wire was a popular choice for use in the construction of residential structures. Aluminum conductors are used in place of copper ones in this design. Aluminum is an excellent conductor, but it has a propensity for expanding and contracting more than copper does, so working with it might raise certain safety issues despite the fact that metal is a good conductor. Aluminum wire is connected with an increased risk of burning and fire dangers, both of which may be mitigated by using correct installation procedures and making use of specific connections. It is vital to keep in mind that aluminum wire has to be regularly inspected, and any alterations or repairs need to be carried out by an electrician who is fully certified.

6. Underground Wiring

Specifically developed for use in installations in which electrical cables are installed underground, subterranean wiring systems are a kind of electrical infrastructure. This sort of wiring utilizes cables or wires that have been specifically created for the application of direct burial and are equipped with shielding and protective layers. In situations where it would be impractical or undesirable to put electrical cables above ground, such as in outdoor lighting systems, electrical connections to outbuildings, subterranean utility networks, and several other applications, underground wiring is often employed.

7. Fiber Optic Wiring

In order to transfer data in the form of pulses of light, fiber optic wire utilizes very fine strands of plastic or glass fibers. Its primary use is for high-speed data transfer, including links to the internet and many forms of telephony.

8. Low Voltage Wiring

Wiring that is considered to be low voltage is used in electrical systems that function at lower voltage levels, often 50 volts or less. It is often used in functions such as doorbell systems, safety devices, and landscape lighting, amongst others.

9. LAN Wiring (Ethernet)

The process of installing cables and connections to construct a LAN, or local area network, for the purpose of data transmission in offices or private residences is known as "LAN wiring." This process is also sometimes referred to as "Ethernet wiring."

10. Coaxial Cable Wiring

Coaxial cables, which have a center conductor encircled by a layer of insulation, a metallic barrier, and an outer shielding layer, are used in coaxial cable wiring. Broadband internet connections and cable television (CATV) are two frequent uses for it.

The many wiring systems utilized in homes and businesses are only a few examples. The right wiring system should be chosen based on the building's age, local electrical regulations, application, and the necessary degree of protection.

Chapter 4: Basics of Wiring

Wiring is the basic act of connecting electromagnetic components and devices inside a circuit in order to permit the passage of electric current. The wiring may also be thought of as the "glue" that holds an electrical system together. It is essential to any electrical system because it enables the transfer of electricity, which is required for the operation of a wide variety of electrical products and systems.

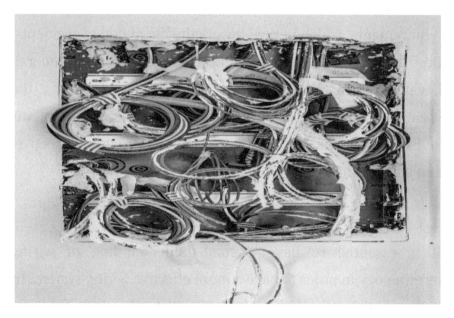

It is essential for electrical experts as well as laypeople to have a fundamental understanding of wiring since this knowledge enables the reliable and effective operation of electrical systems. Learning the vocabulary of wiring is crucial because it offers a uniform language that permits successful communication among electricians, engineers, and others working with electrical structures. This makes learning the jargon of wiring an essential part of the wiring process. Individuals are able to grasp electrical ideas, correctly construct or change circuits, and effectively troubleshoot problems when they have a working knowledge of words such as voltage, resistance, current, conductor, circuit, and insulator, amongst others. In addition, having a solid understanding of the terminology used in wiring is essential for ensuring compliance with electrical rules and safety requirements, which in turn lowers the likelihood of accidents, fires caused by electricity, and electrocutions. In general, having a firm grasp of the foundations as well as the vocabulary of wiring allows one to handle electrical systems and constructions in a manner that is safer, more effective, and better educated.

4.1 Wiring terminology and symbols

It is essential to have a solid understanding of wiring terminology in order to have a more effective wiring system. In light of this, the following is a list of some frequent terminology linked to wiring, as well as explanations of what each phrase means:

Capacitor: A capacitor is a kind of electronic component that is non-active and is used for storing and dispensing electrical energy. It is made up of two plates of conductivity that are separated by a dielectric, which is a kind of substance that provides insulation.

Inductor: When electrical current is sent through an inductor, the component stores potential energy in the shape of a magnetic field. Inductors are classified as passive electronic components. Because it is resistant to variations in current, it is often used in transformers and filters.

Diode: A diode is a kind of semiconductor device that only lets current flow in one direction when it is turned on. In the process of converting alternating current, or AC, to direct current (DC), it is often used as a rectifier.

Transistor: A semiconductor device that can magnify electronic signals, as well as switch between them, is called a transistor. It is equipped with three terminals, which are referred to as the collector, the foundation, and the emitter.

Resistor: A passive electrical part known as a resistor controls the amount of current that may flow through a circuit. It is frequently employed for controlling the amounts of current and the division of voltage.

Ohm's Law: Ohm's Law describes the connection that exists in a circuit among the variables of voltage (V), current (I), and resistance (R), stating that $V = I * R$. In the process of analyzing and designing electrical circuits, one must adhere to this essential premise.

Wattage: Wattage is a unit of measurement used to express the rate at which energy is either consumed or produced in a system of electricity. Watts (W) is used to express it as a measurement since it is a combination of voltage and current.

Joule: The Worldwide System of Numbers uses the joule as its standard unit of measurement for energy. It is used to measure the amount of electrical energy that is generated or consumed over a period of time.

Alternating Current (AC): Alternating current, or AC, is a kind of electric current that changes its direction at regular intervals. The distribution of electricity is one of its most prevalent applications, and its waveform is characterized as sinusoidal.

Direct Current (DC): Direct current, sometimes known as DC, is a form of electricity that exclusively travels in one direction. It is often found in batteries as well as in many electrical equipment.

Ground Fault Circuit Interrupter (GFCI): A ground fault circuit interrupter, or GFCI, is a kind of safety device that helps prevent electric shocks. It does this by monitoring the disparity in current that flows between the hot wires and the neutral wires, and if there is an imbalance, it will trip the circuit.

National Electrical Code (NEC): In the United States, the installation and usage of electrical systems are governed by a set of rules and regulations known as the National Electrical Code (NEC).

Ampacity: Ampacity is the highest amount of current that a conductor is capable of carrying at one time. It is of the utmost importance to make certain that conductors are capable of withstanding the anticipated current load without becoming overheated.

Conduit: In the context of electrical wiring systems, a conduit is a protected tube or channel that is used to house electricity cables and wires. Conduits provide both physical protection and organizational benefits.

Junction box: An electrical splice box or junction panel is a container that is used for the purpose of protecting electrical wires and splices. It not only provides safety but also makes it easier to maintain the electrical system in the future.

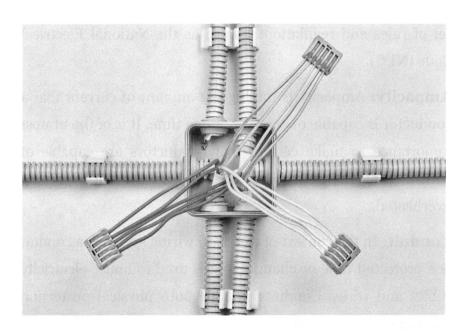

Busbar: Within a wiring panel or switchboard, the distribution of electrical power to many circuits is accomplished with the help of a component known as a busbar, which is a thick and stiff conductor.

Voltmeter: Voltmeters are instruments that are used to measure the amount of voltage that is present across an electrical component or circuit.

Ammeter: A device known as an ammeter is used to determine the amount of electrical current that is moving through a certain component or circuit.

Grounding Electrode: A conducting rod or plate is what we refer to as a grounding electrode. Its purpose is to provide a stable connection between a system of electricity and the ground on Earth.

Insulation Resistance: Insulation resistance measures the ability of an insulating material to prevent current leakage in a circuit.

Load: In the context of an electrical circuit, the term "load" refers to any item or component that draws power from the circuit, such as a lamp, a motor, or an appliance.

Three-phase system: A three-phase system is a type of electrical power distribution that uses three alternating currents, typically used in industrial and commercial applications.

Knockout: A knockout is a removable portion in an electrical box or panel, allowing entry for wiring or conduit.

Conductor Ampacity: Conductor ampacity is the maximum amount of current a conductor can carry safely without exceeding its temperature rating.

Short-circuit current: Short-circuit current is the maximum current that can flow through a circuit when a short circuit occurs.

Load center: A load center, also known as a distribution board or breaker panel, is an electrical panel used to distribute power and protect circuits.

Pull box: A pull box is an enclosure used to facilitate the pulling of wires and cables during installations or repairs.

Polarized Plug: A polarized plug has one prong wider than the other, ensuring the correct connection to a polarized outlet for safety purposes.

Understanding these terminologies is essential for anyone working with electrical systems.

4.2 Wire sizes and gauge selection

Choosing the suitable diameter or thickness of electrical conductors for a given application is one of the most important but often overlooked parts of wiring. This may be accomplished by selecting the appropriate wire size and gauge. An electrical circuit's current-carrying capability, the amount of voltage drop, and the level of overall safety are all directly influenced by the wire size.

The standard units for measuring the size of wire are the American Wire Gauge (AWG) and the metric system. According to the AWG system, bigger wires are given a lower number, while thinner wires are given a higher number. For illustration purposes, a wire with 12 AWG is greater in diameter than a wire with 16 AWG.

The choice of wire gauge is determined by a number of variables, including the following:

Voltage drop: Longer wire runs, or higher currents can result in voltage drops, causing a decrease in voltage at the load end. Properly sized wires minimize this voltage drop to ensure the efficient performance of connected devices.

Temperature rating: The wire should have an appropriate temperature rating to handle the expected operating conditions without deteriorating or melting.

Insulation type: Different applications may require specific types of insulation, such as PVC, thermoplastic, or rubber, depending on factors like moisture exposure, heat, or flexibility requirements.

Overcurrent protection: The wire gauge must be compatible with the overcurrent protection device (e.g., fuse or circuit breaker) used to protect the circuit.

Ampacity: The ampacity of a wire refers to its maximum current-carrying capacity. It should be higher than the maximum current the circuit is expected to carry.

Larger wire gauges, denoted by lower AWG numbers, are appropriate for greater currents and longer runs. On the other hand, smaller wire gauges, denoted by higher AWG numbers, are used for lower currents and shorter distances. Lighting circuits typically utilize 14 AWG wires, whereas general-purpose circuits often use 12 AWG wires. Common wire gauges used in residential applications include 14 AWG and 12 AWG.

Because these standards specify precise criteria for various situations, it is essential to reference the National Electrical Code (NEC) or local electrical rules and recommendations in order to identify the right wire size for a particular installation. By eliminating the risks posed by possible dangers caused by undersized or oversized wires, using the appropriate wire size helps to guarantee that an electrical system is both safe and dependable.

4.3 Wiring color codes

Wiring color codes are defined color schemes that are used in the process of identifying the various conductors that are included in electrical cables and wiring systems. The color codes used in different nations and areas are different; however, there are several colors that are consistently used everywhere. It is very necessary to have a solid understanding of these color codes in order to successfully install, diagnose, and maintain electrical systems. The following is a list of the most frequent color codes used in electrical wiring:

- **Live (Line) Wire**

In most countries, the live or line wire is indicated by the color brown or red. It carries the current from the power source to the load, such as electrical appliances or devices.

- **Neutral Wire**

The neutral wire allows the current to return to the power source. It is typically represented by the color blue in many countries.

- **Earth (Ground) Wire**

The earth or ground wire is used for safety purposes to provide a path for excess electrical current to the earth in the event of a fault. It helps prevent electric shocks and fires. The color green with a yellow stripe (green/yellow) is commonly used to designate the earth wire.

- **Three-Phase Wiring**

In three-phase electrical systems, additional color codes are used to identify each phase conductor. In many regions, the phases are identified as follows:

- Phase 1: Brown
- Phase 2: Black
- Phase 3: Gray

- **Switched Live Wire**

This wire is used in two-way and intermediate switching circuits to control the power supply to a device. The color brown or red is often used for switched live wires.

- **Control or Signal Wires**

In low-voltage control circuits, different colors may be used to represent specific signals or functions. For example:

> ➤ Control voltage (24V, 12V, etc.): Blue, Orange, or Violet

> ➤ Signal or communication wires: Gray, White, or Purple

It is very important to keep in mind that the color codes described above are broad norms and that the precise color codes used may differ from one nation or location to another. In addition, some older installations could not adhere to the

current color codes. Because of this, it is essential to utilize extra procedures, such as labeling or testing, to verify that the wires are correctly identified. It is essential to adhere to wiring color standards in order to guarantee both the safety of electrical systems and their ability to function effectively. When dealing with electrical wiring, it is critical to constantly double-check the neighborhood norms and standards to ensure that your work is consistent and in accordance with electrical regulations.

Chapter 5: Wiring Techniques and Practices

Before beginning real wiring, it is necessary to acquire the skills and knowledge necessary for wiring in order to assure both safety and efficiency. Accidents, electric shocks, and even fires may all be avoided by learning how to properly handle wires, establishing safe connections, and adhering to all applicable electrical rules. Installations that are trustworthy are also the result of using the appropriate procedures, which reduces the likelihood of malfunctions and the need for expensive repairs. Gaining a solid basis for managing electrical systems and increasing one's level of confidence while dealing with live circuits are both benefits of learning the basics. To summarize, learning the appropriate wiring procedures and best practices in advance is very necessary in order to provide an electrical environment that is risk-free and productive.

5.1 Proper wire stripping and termination

Skill in correctly stripping and terminating wires is essential for employment in the electrical trade. Termination is the process of safely attaching a wire that has been stripped to a terminal or connector. Wire stripping is the process of carefully removing the insulation from the ends of electrical wires in order to expose the conductive metal below. The successful mastery of these skills is necessary for ensuring that electrical connections are both safe and dependable.

1. Wire Stripping

Make use of a wire stripper or a utility knife to ensure that the wire is striped properly. The insulation must be stripped away without causing any harm to the conductive core in order to achieve the goal. The following are the stages:

- **Measure**

Find out how much wire is required for the termination point, and then add a little bit more so there will be room for changes.

- **Select the right tool:**

If you do not have access to a wire stripper, you may either use a utility knife or a wire stripper that corresponds to the wire gauge.

- **Position the wire**

Position the wire so that it is sandwiched between the right size stripper or knife, taking care not to apply an excessive amount of pressure.

- **Strip the insulation**

To remove insulation from the wire, either slowly spin the stripper or make a controlled cut around the wire's perimeter with the knife. Take care not to cut too deeply in order to avoid doing any harm to the conductor.

- **Remove the insulation**

Following the completion of the cut, remove the insulation from the wire by carefully pulling it away from the conductor. Make sure that none of the wire strands are severed or otherwise harmed throughout the operation.

2. **Wire Termination**

For a wire to be terminated, it must first be stripped and then securely connected to termination or connection. A secure and stable connection may be achieved by proper termination, which also eliminates the risk of connections becoming loose or unstable. The following are some of the most frequent methods:

- **Screw Terminal**

In this approach, the wire that has been stripped is wrapped around a screw, and the screw is then tightly tightened. Make sure the wire is wrapped around the screw in a clockwise direction to prevent it from unraveling while you are tightening it.

- **Crimping**

Crimp terminals need a specific crimping tool in order to create a secure connection between the wire and the terminal by compressing the terminal around the wire. The crimp should have enough force to securely grab the wire, but it shouldn't have so much force that it damages the wire.

- **Soldering**

Soldering, also known as soldering, is a process that involves melting a metal alloy to produce a strong connection between a wire and a terminal. It has great conductivity and helps protect the connection from corrosion at the same time.

- **Push-In or Quick Connect Terminals**

There are certain terminals that do not need the wires to be stripped before they may be plugged straight into them. These terminals often feature devices that are loaded with springs in order to hold the wire.

- **Insulation**

Following the termination, check to see that there are no exposed conductors remaining. To insulate and safeguard the connection, you may shield it using electrical tape, heat-shrink tubing, or any other way that is suitable.

Wires need to be stripped and terminated correctly in order to avoid loose connections, lower the danger of electrical risks, and keep the integrity of electrical systems intact.

5.2 Basic wiring connections (e.g., switches, outlets, lights)

The following is a condensed description of the fundamental connections in electrical wiring for typical electrical components such as switches, outlets, lights, and other items:

1. **Switches**

Single-Pole Switch: This is the most popular form of switch used to control lights or other devices from a single place. It consists of a single pole. The hot wire, which is often black in color, is connected to one of the screw terminals, and the switched hot wire, which is typically red in color, is connected to the other terminal.

Three-Way Switch: A light or other device may be controlled from two distinct places with the help of a three-way switch. They are equipped with three screw terminals: one common terminal, which is often colored black, and two traveler terminals, which are typically colored brass or silver.

2. **Outlets (Receptacles)**

Duplex Outlet: The typical domestic outlet that has two openings for several types of plugs. It has two terminals made of brass for the hot wires, two terminals made of silver for the neutral wires, and one termination made of green for the ground wire.

Ground Fault Circuit Interrupter (GFCI) Outlet: Ground Fault Circuit Interrupter (GFCI) outlets were developed to prevent electrical shocks. They are equipped with "Test" and "Reset" buttons, and their operation is based on detecting imbalances in the flow of electricity.

3. Lights

Light Fixture: Light fixtures feature two or more wires (hot, neutral, and often a ground wire), depending on the number of bulbs in the fixture. The white wire serves as the connection point for the neutral, while the black or red wire of the fixture serves as the connection point for the hot wire. The green or bare wire on the fixture is where the ground wire is supposed to join.

4. Three-Phase Connections

Motor with Three Phases: Motors used in three-phase systems have three power lines (L1, L2, and L3) in addition to a ground connection. Each lead supplies a different phase voltage, and the ground prevents faults from occurring.

Three-Phase Panel: A three-phase electrical panel will include a neutral busbar in addition to the three hot busbars that are designated for each phase. Circuit breakers and fuses safeguard the individual circuits by connecting to each phase of the power supply.

5. Electrical Junction Boxes

Wire connections are housed inside junction boxes so that they are protected from being accidentally tripped over.

They occur in a wide variety of shapes, sizes, and sorts, and they are essential for maintaining order and security.

6. Ground Connections

Grounding Rod: A grounding rod is a metal rod that is pushed into the ground at the electrical service entry of a building in order to establish a dependable ground connection.

Grounding Clamp: A grounding clamp is used to attach the grounding conductor to the grounding electrode (for example, a rod or a metal water pipe) so that surplus current has a route that is safe to go.

7. Wire Splices

Wire Nuts: Wire nuts are twist-on connectors that are used in the process of securely joining many wires together.

Crimp Connectors: Crimp connectors include the use of a crimping tool in order to squeeze the wires together and secure them.

Keep in mind that any work involving electricity must always be carried out with caution and in accordance with the local electrical rules and safety requirements.

5.3 Grounding and bonding principles

When it comes to electrical wiring, grounding and bonding are two fundamental elements that are necessary to assure safety and provide protection against electrical risks. Both procedures entail creating a route with low resistance through which

electricity may flow in the event of an electrical defect, such as a short circuit or a malfunction in a piece of equipment. Let's look at a more in-depth explanation of bonding and grounding:

- **Grounding**

In order to provide a route that is free from the danger of fault currents, grounding entails connecting electrical systems and equipment to the surface of the earth. The following are the primary goals of grounding:

- **Protection against Electric Shocks**

In the case of a problem, such as a short circuit, grounding will route the fault current away from people and equipment and toward the Earth, which will reduce the danger of electric shock.

- **Fault Current Detection**

A reference point is provided via grounding, which enables fault currents to be detected. When an error occurs, it causes an imbalance in the electrical system, which in turn activates protection mechanisms such as circuit breakers, which cut power to the circuit that has the problem.

- **Equipment Protection**

By ensuring that fault currents have a channel to flow that has a low resistance, grounding serves to safeguard electrical equipment by lowering the risk of possible damage as well as fire threats.

The following are examples of common grounding methods:

Grounding Electrode System: Linking electrical systems to one or more grounding electrodes that are in direct touch with the Earth, such as grounding rods or metal water pipes, is required to accomplish this.

Equipment Grounding: The metal casings and enclosures of electrical devices are connected to the grounding system through equipment grounding. This creates a conduit for fault currents to travel along.

Grounding Conductors: Grounding connections often make use of insulated conductors that are either green or green/yellow in color.

Bonding: In an electrical system, the process of joining metallic components and enclosures in order to produce electrical continuity and equipotential bonding is referred to as bonding. The primary goals of the bonding process are:

Eliminate Voltage Potential: Bonding brings together all of the metallic components of an assembly, such as conduit systems, enclosures, and metal frames, so that they share the same voltage potential. This eliminates the potential for voltage variations to cause injuries from electric shock or to cause damage to equipment.

Reduce Electromagnetic Interference: Bonding helps decrease electromagnetic interference by avoiding circulating currents, which may produce noise in sensitive electronic systems. This is accomplished by providing low-impedance routes for currents to pass.

Ground Fault Current Path: Bonding that is done correctly will produce an efficient ground fault current channel, which will enable fault currents to flow swiftly and securely. This will guarantee that protective devices will activate fast when they are needed.

Static Electricity Discharge: Bonding helps eliminate static charges, which lowers the likelihood of electrical sparks occurring in potentially dangerous conditions.

Bonding may be accomplished by connecting all metallic components together via bonding jumpers, bonding conductors, or metal conduits. This creates an electrical connection.

Both grounding and bonding are very important in terms of ensuring the safety of electrical systems and should be performed in accordance with the electrical regulations and standards that are particular to each location. Electrical systems that are properly grounded and bonded provide safety for people, equipment, and property. This protects against the possibility of electrical mishaps and ensures that electrical installations continue to function in a smooth and dependable manner.

Chapter 6: Wiring Projects

To get a comprehensive understanding of wiring projects, one must take a methodical approach and draw upon both their theoretical background and their hands-on, practical expertise. To assist you in gaining a comprehensive understanding of wiring projects, the following is a step-by-step guide:

Basic Electrical Theory: Learn the basic ideas of electricity first, such as voltage, current, resistance, and Ohm's law. After that, go on to more advanced topics. Gain an understanding of how electrical circuits operate and the many components that make up electrical systems.

Study Wiring Codes and Regulations: Make sure you are familiar with the electrical rules and regulations that are in effect in your location. In electrical installations, compliance with codes is required to assure both safety and consistency.

Safety Precautions: Always put your own safety first. Gain an understanding of the best practices for electrical safety, such as the necessity of wearing personal protective equipment (PPE), maintaining safe working distances, and turning off the power before beginning any electrical work.

Study Wiring Diagrams: Exercise your reading and comprehension skills by working through several wiring diagrams. You will be better able to envision how components are linked together with the assistance of these diagrams, which illustrate the architecture and connections of electrical circuits.

Hands-On Training: Either enroll in a structured electrical training course or work on some basic wiring projects under the direction of an experienced mentor in order to get some practical expertise in the field.

Start Simple: Start with simple chores like wiring a single-pole switch, installing an electrical outlet, or attaching a light fixture. These are all good places to get your feet wet. As your self-assurance grows, you should challenge yourself with more difficult tasks.

Plan and Prepare: Before beginning a project, it is important to make a detailed design for the wiring layout and collect all of the required supplies as well as equipment. Errors are cut down on, and time is saved thanks to proper preparation.

Proper Wire Stripping and Termination: Acquire the skills of wire stripping and termination in order to construct connections that are safe and dependable.

Practice Wiring Techniques: Acquire knowledge of a variety of wiring approaches, including parallel and series connections, and put this knowledge into effect with a variety of projects.

Troubleshooting: Acquire the abilities necessary for troubleshooting in order to locate and fix problems with the wiring, such as open circuits, short circuits, or defective connections.

Document and Label: Label and record your wiring projects in the correct manner, as this will assist with any future maintenance or troubleshooting that may be required.

Take on Diverse Projects: In order to get more expertise, you should get your hands dirty with a wide range of wiring tasks, such as residential, commercial, and industrial applications.

Stay Updated: Maintain an awareness of the latest developments in electrical technology as well as any changes that may occur in electrical rules and regulations.

Get Certified: Consider obtaining the necessary certificates or licenses in the field of electrical work since these may indicate your level of skill and trustworthiness to prospective customers or companies. Keep in mind that in order to master electrical wiring tasks, you will need patience, practice, and a commitment to ongoing progress.

6.1 Installing Switches and Outlets for Wiring

The procedures and fundamentals involved in installing switches and outlets are almost identical. An explanation broken down into steps is provided below for installing a switch in addition to an electrical outlet:

Note: For your own protection and the protection of others, always switch off the power to the circuit at the circuit breaker or fuse box before beginning any kind of electrical work. Utilize a voltage tester to confirm that there is no flow of power to the switch or outlet in question.

Materials and Tools Needed

- Switch or outlet (new)
- Screwdriver
- Wire stripper (if needed)
- Voltage tester
- Wire nuts (for connections)

Switch Installing

Step 1: Prepare for the Project

- Cut off the energy to the circuit at the circuit breaker or fuse box.

- Use a voltage meter to check the cables in the existing switch box to make sure there is no electrical current.

Step 2: Destroy the Old Switch (if applicable)

- Unscrew the switch plate cover and remove it.
- Unscrew the switch from the electrical box using a screwdriver.
- Carefully pull the switch out from the box, but do not disconnect any wires just yet.

Step 3: Identify the Wires

- Examine the wires connected to the old switch.
- There are typically three wires:

1. Warm wire, often in red or black - This wire carries the electrical current from the power source.

2. Switched hot wire (usually black or red) - This wire carries the current to the light fixture when the switch is on.

3. Ground wire (usually green or bare copper) - This wire provides a path for fault currents to prevent electric shocks.

Step 4: Disconnect the Old Switch

- Unscrew the wire nuts that connect the wires to the old switch.
- Carefully remove the old switch from the wires.

Step 5: Prepare the New Switch

- If necessary, strip the ends of the wires using a wire stripper to expose about 1/2 inch of copper.
- Most switches have screw terminals labeled "Common," "Line," and "Ground."

Step 6: Connect the Wires to the New Switch

- Connect the hot wire to the "Line" terminal of the new switch.
- Connect the switched hot wire to the "Common" terminal of the new switch.
- Connect the ground wire to the "Ground" terminal of the new switch.
- Tighten the screws securely to ensure proper connections.

Step 7: Mount the New Switch

- Gently push the wires and the new switch into the electrical box.
- Secure the new switch to the box with the screws provided.

Step 8: Attach the Switch Plate Cover

- Screw the switch plate cover over the new switch.

Step 9: Turn on the Power

- Go back to the circuit breaker or fuse box and turn on the power to the circuit.

Step 10: Test the Switch

- Test the new switch by turning it on and off to ensure it controls the light fixture correctly.

Installing an Electrical Outlet

Step 1: Prepare for the Project

- Turn off the power to the circuit at the circuit breaker or fuse box.

- Test the wires in the existing outlet box with a voltage tester to ensure there is no electrical current.

Step 2: Remove the Old Outlet (if applicable)

- Unscrew the outlet plate cover and remove it.

- Unscrew the outlet from the electrical box using a screwdriver.

- Carefully pull the outlet out from the box, but do not disconnect any wires just yet.

Step 3: Identify the Wires

- Examine the wires connected to the old outlet.

- There are typically three wires:

1. The warm wire, which is often black or red, is used to transmit electrical current from the power source.

2. Balanced wire, which is often white, is the wire that returns current to the power source.

3. The ground wire, which is often green or bare copper, serves as a conduit for fault currents to avoid electric shocks.

Step 4: Disconnect the Old Outlet

- Unscrew the wire nuts that connect the wires to the old outlet.
- Carefully remove the old outlet from the wires.

Step 5: Prepare the New Outlet

- If necessary, strip the ends of the wires using a wire stripper to expose about 1/2 inch of copper.
- Most outlets have screw terminals labeled "Line," "Load," "Neutral," and "Ground."

Step 6: Connect the Wires to the New Outlet

- Connect the hot wire to the "Line" terminal of the new outlet.
- Connect the neutral wire to the "Neutral" terminal of the new outlet.

- Connect the ground wire to the "Ground" terminal of the new outlet.

- Tighten the screws securely to ensure proper connections.

Step 7: Mount the New Outlet

- Gently push the wires and the new outlet into the electrical box.

- Secure the new outlet to the box with the screws provided.

Step 8: Attach the Outlet Plate Cover

- Screw the outlet plate cover over the new outlet.

Step 9: Turn on the Power

- Go back to the circuit breaker or fuse box and turn on the power to the circuit.

Step 10: Test the Outlet

- Test the new outlet by plugging in a device to ensure it receives power correctly.

That's it! You have successfully installed a light switch and an electrical outlet.

6.2 Wiring light fixtures and ceiling fans

Wiring light fixtures and ceiling fans involve similar principles as other wiring projects. Here's a step-by-step explanation of this wiring project:

Note: Before starting any electrical work, turn off the power to the circuit at the circuit breaker or fuse box to ensure safety. Use a voltage tester to verify that there is no electricity flowing to the wires.

Materials and Tools Needed

- Light fixture and ceiling fan with mounting hardware
- Screwdriver (flathead or Phillips, depending on the screw type)
- Voltage tester
- Wire stripper (if needed)
- Wire nuts (for connections)
- Ladder (if required to reach the ceiling)

Wiring Light Fixtures

Step 1: Prepare for the Project

- Turn off the power to the circuit at the circuit breaker or fuse box.

- Test the wires in the existing light fixture box with a voltage tester to ensure there is no electrical current.

Step 2: Remove the Old Fixture (if applicable)

- Unscrew and remove the existing light fixture from the ceiling.

- Disconnect the wires from the old fixture and set it aside.

Step 3: Identify the Wires

- Examine the wires in the electrical box in the ceiling.

- There are typically three wires:

1. Hotwire (usually black or red) - This wire carries the electrical current from the power source.

2. Neutral wire (usually white) - This wire carries the current back to the power source.

3. Ground wire (usually green or bare copper) - This wire provides a path for fault currents to prevent electric shocks.

Step 4: Prepare the New Light Fixture

- If necessary, strip the ends of the wires using a wire stripper to expose about 1/2 inch of copper.

- The light fixture will have wires for hot, neutral, and ground.

Step 5: Connect the Wires to the New Light Fixture

- Connect the hot wire from the ceiling to the hot wire of the light fixture.

- Connect the neutral wire from the ceiling to the neutral wire of the light fixture.

- Connect the ground wire from the ceiling to the ground wire of the light fixture.

- Secure the wire connections with wire nuts and ensure they are tight and well-insulated.

Step 6: Mount the New Light Fixture

- Follow the manufacturer's instructions to mount the new light fixture to the ceiling box using the provided mounting hardware.

Step 7: Attach the Light Fixture Cover

- Place the light fixture cover over the fixture and secure it with screws or clips provided.

Step 8: Turn on the Power

- Go back to the circuit breaker or fuse box and turn on the power to the circuit.

Step 9: Test the Light Fixture

- Test the new light fixture by flipping the wall switch to ensure it illuminates correctly.

Wiring Ceiling Fans with Light Fixtures

Step 1: Prepare for the Project

- Turn off the power to the circuit at the circuit breaker or fuse box.

- Test the wires in the existing ceiling fan box with a voltage tester to ensure there is no electrical current.

Step 2: Remove the Old Ceiling Fan (if applicable)

- Unscrew and remove the existing ceiling fan from the ceiling.

- Disconnect the wires from the old fan and set it aside.

Step 3: Identify the Wires

- Examine the wires in the electrical box in the ceiling.

- There are typically three wires for the light fixture:

1. Hotwire (usually black or red) - This wire carries the electrical current from the power source.

2. Neutral wire (usually white) - This wire carries the current back to the power source.

3. Ground wire (usually green or bare copper) - This wire provides a path for fault currents to prevent electric shocks.

Step 4: Prepare the New Ceiling Fan

- If necessary, strip the ends of the wires using a wire stripper to expose about 1/2 inch of copper.

- The ceiling fan will have wires for hot, neutral, and ground for both the fan and the light kit.

Step 5: Connect the Wires to the Ceiling Fan

- Connect the hot wire from the ceiling to the hot wire of the ceiling fan for the fan function.

- Connect the neutral wire from the ceiling to the neutral wire of the ceiling fan for the fan function.

- Connect the ground wire from the ceiling to the ground wire of the ceiling fan for the fan function.

- Connect the hot wire from the ceiling to the hot wire of the ceiling fan for the light function.

- Connect the neutral wire from the ceiling to the neutral wire of the ceiling fan for the light function.

- Connect the ground wire from the ceiling to the ground wire of the ceiling fan for the light function.

- Secure the wire connections with wire nuts and ensure they are tight and well-insulated.

Step 6: Mount the New Ceiling Fan

- Follow the manufacturer's instructions to mount the new ceiling fan to the ceiling box using the provided mounting hardware.

Step 7: Attach the Fan Blades and Light Fixture Cover

- Attach the fan blades and the light fixture cover following the manufacturer's instructions.

Step 8: Turn on the Power

- Go back to the circuit breaker or fuse box and turn on the power to the circuit.

Step 9: Test the Ceiling Fan and Light Fixture

- Test the ceiling fan by operating the fan function and the light function using the provided wall switches or remote control.

That's it! You have successfully wired light fixtures and a ceiling fan with light.

6.3 Extending or adding new circuits

Extending or adding new circuits is a more complex electrical wiring project that involves running new electrical cables to power additional outlets, switches, or appliances. Here's a step-by-step explanation of this wiring project:

Note: This project requires a good understanding of electrical principles, and it is advisable to consult a qualified electrician or obtain proper permits before attempting this project. Always turn off the power to the circuit at the circuit breaker or fuse box before starting any electrical work.

Materials and Tools Needed

- Electrical cables of appropriate gauge and type (determined by the circuit's amperage and the intended use)
- Junction boxes
- Outlet boxes or switch boxes
- Circuit breaker
- Wire nuts

- Screwdriver (flathead or Phillips, depending on the screw type)
- Wire stripper
- Fish tape or wire snake (for pulling cables through walls or ceilings)
- Hammer and nails or screws (for securing boxes)
- Voltage tester

Step 1: Plan the Circuit Extension

- Determine the new locations for outlets, switches, or appliances, and plan the routing of the new electrical cables.
- Calculate the total load of the new circuit to ensure it does not exceed the capacity of the circuit breaker and the wire gauge.

Step 2: Turn Off the Power

- Turn off the power to the circuit at the circuit breaker or fuse box that you will be extending.

Step 3: Run the New Electrical Cables

- Carefully run the new electrical cables through walls, ceilings, or along surfaces using fish tape or wire snakes.

- Ensure the cables are protected and not exposed to potential hazards.

Step 4: Install Junction Boxes and Outlet Boxes/Switch Boxes

- Install junction boxes where splices or connections will be made for the new circuit.

- Install outlet boxes or switch boxes at the desired locations for outlets or switches.

Step 5: Connect the New Cables to the Existing Wiring

- Strip the ends of the new cables and existing wiring using a wire stripper to expose about 1/2 inch of copper.

- Use wire nuts to securely connect the corresponding wires: hot to hot, neutral to neutral, and ground to ground.

Step 6: Secure and Mount the Boxes

- Securely mount the junction boxes, outlet boxes, and switch boxes to the walls or ceilings using nails or screws.

- Ensure the boxes are flush with the surface and do not protrude.

Step 7: Connect New Circuit to the Circuit Breaker

- Install a new circuit breaker in the circuit breaker panel of appropriate amperage to match the total load of the new circuit.

- Connect the hot wire (black or red) of the new circuit to the new circuit breaker.

- Connect the neutral wire (white) of the new circuit to the neutral bus bar in the panel.

- Connect the ground wire (green or bare copper) of the new circuit to the grounding bus bar in the panel.

Step 8: Complete the Wiring

- Finish connecting all the wires in the junction boxes, outlet boxes, and switch boxes according to your wiring plan.

- Double-check all connections to ensure they are secure and correctly terminated.

Step 9: Test the Circuit

- Turn on the power to the circuit at the circuit breaker or fuse box.

- Test the new circuit by operating the outlets, switches, or appliances connected to it.

Step 10: Inspect and Secure the Wiring

- Conduct a thorough inspection of all wiring connections, junction boxes, and outlet boxes to ensure they are up to code and meet safety standards.

- Use cable staples or other approved methods to secure the new electrical cables and prevent them from hanging loosely.

Chapter 7: Troubleshooting and Maintenance for Wiring

Troubleshooting: It is the process of locating and fixing errors, flaws, or problems that crop up inside a system. It entails using methodical approaches to problem-solving in order to identify the fundamental reason behind a malfunction or breakdown and then putting into action suitable remedies in order to resolve the problem. In a wide variety of contexts, including technology, electronics, mechanical systems, software, and electrical systems, troubleshooting is a technique that is often used. In the context of electrical systems, the term "troubleshooting" refers to the process of locating and resolving issues that may manifest in the electrical wiring, circuits, devices, or appliances that make up the system. Common electrical troubleshooting scenarios include the detection of faulty wiring, the diagnosis of circuit overloads, the localization of power outages' causes, the resolving of switch and outlet issues, and the repair of electrical equipment that is not functioning properly.

Maintenance: on the other hand, is the practice of providing routine care and upkeep to a system or piece of equipment in order to ensure that it operates correctly, avoids breakdowns, and prolongs its lifetime. Depending on the kind of system, the actions that constitute maintenance could seem somewhat different from one another, but in general, they consist of checking, cleaning, lubricating, adjusting, and replacing components that have been damaged or worn out. In the context of electrical systems, maintenance entails performing routine inspections and taking corrective measures in order to guarantee the integrity of the electrical wiring, circuits, devices, and equipment, as well as their capacity to function safely and effectively. The testing of outlets and switches, the cleaning and tightening of connections, the inspection for indications of wear on electrical components, the replacement of defective components, and the elimination of possible risks are all included in routine maintenance.

Both troubleshooting and maintenance are procedures that are interrelated and mutually beneficial to one another. Maintenance helps stop problems from occurring in the first place by proactively resolving possible concerns and ensuring that the system is kept in excellent working order. This reduces the likelihood of problems occurring. However, even with routine maintenance, there is always a possibility that a malfunction or breakdown may take place at some point. In these cases, troubleshooting will be necessary. It is vital to do troubleshooting in order to identify and fix unforeseen problems, as well as to return the system to its best operational condition. Electrical systems, whether they are used in residential, commercial, or industrial settings, all need regular maintenance and troubleshooting as critical activities to assure their continued dependability, safety, and overall effectiveness. Troubleshooting gives the required ability to detect and handle problems as quickly as they develop, while routine maintenance lowers the chance of significant failures and extends the life of electrical components.

7.1 Identifying and resolving common wiring issues

It is essential to both recognize and fix the usual wiring problems that arise in order to keep an electrical system operational and risk-free. The following is a list of the most frequent wiring problems and the solutions to those problems:

1. **Flickering Lights**

    • Check if the lightbulb is securely screwed in and not loose.

    • Replace the lightbulb with a new one to see if the issue persists.

    • If the problem continues, there might be a loose connection in the circuit or a faulty switch. Turn off the power and check and tighten all connections.

2. **Tripping Circuit Breakers or Blown Fuses:**

- This could indicate an overloaded circuit. Unplug or turn off some devices to reduce the load.

- If the problem persists, there might be a short circuit or ground fault. Inspect outlets, switches, and devices for damage and exposed wires.

3. **Dead Outlets or Switches:**

- Check the circuit breaker or fuse box to see if the corresponding circuit has tripped or blown a fuse.

- If the circuit breaker is on, turn off the power and inspect the outlet or switch for loose or damaged wires. Reconnect or replace as needed.

4. **Warm or Hot Outlets or Switches:**

- Warm outlets or switches could indicate an overloaded circuit or a loose connection.

- Unplug devices or turn off switches on that circuit and check the wiring for loose connections. Consider redistributing the load if necessary.

5. **Sparks or Electrical Arcing:**

- Sparks or electrical arcing can be dangerous and indicate a serious issue.

- Turn off the power immediately and inspect the outlet or switch for damaged or loose wires. If you find any, replace the outlet or switch.

6. **GFCI Tripping:**

    • Ground Fault Circuit Interrupters (GFCIs) protect against electrical shocks and may trip due to ground faults.

    • Press the "Reset" button on the GFCI to restore power. If it keeps tripping, there might be a fault in the connected device or wiring.

7. **High Electricity Bills:**

    • High electricity bills could result from inefficient appliances, improper insulation, or power leaks.

    • Inspect appliances for energy efficiency and consider upgrading to more energy-efficient models. Ensure doors and windows are properly sealed.

8. **Open or Cut Wires:**

    • If you discover open or cut wires, immediately turn off the power and repair or replace the damaged section.

    • Use appropriate electrical tape, wire nuts, or soldering to reconnect the wires securely.

9. **Rodent Damage:**

- Rodents may chew on wires, leading to dangerous conditions.

- Inspect areas where rodents may have access and address any damage found. Consider using rodent-proofing measures.

10. **Improperly Wired Outlets or Switches:**

    - If an outlet or switch does not function as expected, it might be improperly wired.

    - Double-check the connections and wiring diagram for the specific device and correct any mistakes.

It's important to prioritize safety and turn off the power when troubleshooting or resolving any wiring issues.

7.2 Regular maintenance and safety checks

Regular maintenance and safety checks are crucial for wiring to ensure the safety, efficiency, and reliability of the electrical system. Here are the key reasons why regular maintenance and safety checks are essential for wiring:

- **Prevent Electrical Hazards**

Regular maintenance helps identify and address potential electrical hazards before they escalate into dangerous situations. Electrical issues, such as loose connections, damaged wires, or overloaded circuits, can lead to electrical shocks, fires, or even electrocution. By proactively inspecting and repairing these problems, the risk of accidents and injuries can be significantly reduced.

- **Extend Lifespan of Wiring and Devices**

Regular maintenance helps keep the wiring and electrical components in good condition. Proper care and upkeep can prevent premature wear and tear, corrosion, or degradation of wires, switches, outlets, and other electrical devices. This extends their lifespan and reduces the need for frequent replacements.

- **Ensure Optimal Performance**

Well-maintained wiring and electrical components function optimally. When connections are tight, circuits are balanced, and devices are in good working condition, the electrical system operates efficiently. This can lead to energy savings and better performance of electrical appliances and equipment.

- **Compliance with Electrical Codes**

Regular safety checks ensure that the electrical system meets the relevant electrical codes and safety standards. Compliance with codes is not only a legal requirement but also a critical aspect of maintaining a safe environment for occupants or users of the building.

- **Detect Early Signs of Problems**

Regular inspections can catch early signs of potential issues, allowing for timely repairs and preventing minor problems from escalating into major faults. Addressing minor problems can save time and money in the long run.

- **Reduce Downtime and Disruptions**

Scheduled maintenance helps identify and resolve wiring issues before they cause unexpected failures or outages. This reduces the likelihood of downtime or disruptions to the electrical system, which is especially crucial in commercial and industrial settings where productivity and operations rely heavily on a continuous power supply.

- **Preserve Property and Assets**

Electrical fires or failures caused by faulty wiring can lead to significant property damage and loss of assets. Regular maintenance and safety checks minimize the risk of electrical incidents that could result in property damage, saving repair costs and preserving valuable assets.

- **Peace of Mind**

Knowing that the electrical system is well-maintained and safe provides peace of mind to occupants, homeowners, or building managers. It instills confidence in the reliability and safety of the electrical infrastructure. Regular maintenance and safety checks are essential to ensure the safety, longevity, and optimal performance of electrical wiring and components. A well-maintained electrical system not only reduces the risk of accidents and hazards but also enhances energy efficiency and protects valuable assets.

Chapter 8: Advanced Topics of Wiring

The passage of time has resulted in an improvement in all aspects. In the same manner, wiring is also becoming better with each passing day, which is altering the way people live. In the realm of wiring, there are a few advanced topics:

8.1 Wiring for specialized applications

The term "wiring for specialized applications" refers to electrical wire installations that are conceived of, planned for, and carried out in order to fulfill the particular needs of one-of-a-kind or specialized settings, pieces of equipment, or whole systems. These applications often have specific safety, performance, or regulatory issues that call for particular wiring methods, materials, or configurations. This is necessary in order to meet the demands of these factors.

8.2 Smart home wiring and automation

The term "smart home wiring and automation" refers to the process of incorporating cutting-edge technology and intelligent systems into a residential property in order to improve the level of convenience, comfort, safety, and energy efficiency, as well as the level of entertainment available. This requires the installation of specialized cabling, sensors, intelligent devices, and control systems that can be remotely monitored and handled via the use of mobile devices like smartphones and tablets, as well as voice-activated assistants. The following is an explanation of the wiring and automation involved in a smart home:

1. Wiring Infrastructure

- Smart home wiring starts with a robust infrastructure of structured cabling. This includes running Ethernet cables for high-speed internet connectivity, coaxial cables for cable or satellite TV, and possibly fiber optic cables for even faster data transmission.

- Additionally, low-voltage wiring is essential for powering and connecting various smart devices and sensors throughout the home. These devices may include smart light switches, smart thermostats, motion sensors, door/window sensors, and smart speakers.

2. Smart Devices:

- Smart home devices are internet-connected devices that can be controlled remotely or programmed to perform specific tasks automatically.

- Common smart devices include smart thermostats for heating and cooling control, smart lighting systems that allow remote dimming and scheduling, smart door locks for keyless entry, smart cameras for home security, and smart appliances that can be operated and monitored remotely.

3. Home Automation Systems:

- Home automation systems are central hubs that connect and control various smart devices throughout the home.

- These systems can be managed through mobile apps or voice-activated virtual assistants like Amazon Alexa or Google Assistant.

- They allow users to create custom automation routines and scenes where multiple devices can be triggered to perform specific actions simultaneously.

4. Energy Efficiency:

- Smart home automation can significantly improve energy efficiency. For example, smart thermostats can learn users' preferences and automatically adjust heating and cooling based on occupancy patterns, resulting in reduced energy consumption.

- Smart lighting can be programmed to turn off when rooms are vacant or dimmed during specific hours to save electricity.

5. Security and Surveillance:

- Smart home security systems integrate cameras, motion sensors, and door/window sensors with mobile alerts to enhance home security.

- Users can receive real-time alerts and remotely monitor their property through their smartphones.

6. Entertainment and Audio-Visual Integration:

- Smart home automation extends to entertainment systems, where users can control audio, video, and streaming services from a single interface.

- Home theaters can be integrated with lighting and sound systems, allowing users to create immersive movie-watching experiences with a single command.

7. Voice Control and Integration:

- Many smart home devices are compatible with popular voice-activated virtual assistants, allowing users to control various aspects of their home using voice commands.

- Voice control enhances convenience and accessibility for users, especially when they are not near their smartphones or control panels.

In general, smart home wiring and automation provide a connected and intelligent living environment, which in turn provides occupants with increased ease of use, comfort, and security, in addition to improved energy efficiency.

8.3 Energy-efficient wiring practices

Implementing tactics and utilizing equipment that decreases both the amount of power that is used and wasted are two components of energy-efficient wiring practices. An explanation of energy-efficient wiring procedures is provided in the following step-by-step format:

Step 1: Plan Your Wiring Layout

- Before starting the wiring project, plan the layout carefully to minimize the length of wiring runs.

- Place outlets and switches strategically to avoid excessive use of extension cords or long cables.

Step 2: Use High-Quality Wiring and Components

- Invest in high-quality electrical cables and components that have low resistance and can handle the required loads efficiently.

- Consider using energy-efficient devices and appliances with high Energy Star ratings.

Step 3: Opt for LED Lighting

- Use LED (Light Emitting Diode) bulbs and fixtures for lighting. LEDs are highly energy-efficient, consuming significantly less power than traditional incandescent or fluorescent bulbs.

- LEDs also have a longer lifespan, reducing the frequency of replacements and minimizing waste.

Step 4: Install Dimmer Switches

- Dimmer switches allow you to adjust the brightness of lights according to your needs, saving electricity when full brightness is not required.

- This can extend the lifespan of LED bulbs and reduce energy consumption.

Step 5: Implement Zoning and Smart Controls

- Consider using zoning and smart controls for heating, ventilation, and air conditioning (HVAC) systems.

- Zoning allows you to heat or cool specific areas of your home as needed, avoiding energy wastage in unoccupied spaces.

- Smart controls, such as programmable thermostats, enable you to set temperature schedules to optimize energy usage.

Step 6: Utilize Daylight and Natural Ventilation

- Maximize the use of natural daylight by placing windows strategically and using reflective surfaces to enhance lighting.

- Use natural ventilation whenever possible to reduce the need for artificial cooling and save energy.

Step 7: Apply Power Strips and Unplug Devices

- Use power strips for multiple electronic devices to easily turn them off when not in use, eliminating standby power consumption.

- Unplug chargers, devices, and appliances when not needed, as they can draw power even in standby mode.

Step 8: Insulate Wiring Properly

- Properly insulate wiring to reduce heat loss and energy wastage. This is especially important for high-power electrical equipment and appliances.

Step 9: Regular Maintenance and Inspections

- Schedule regular inspections and maintenance of your electrical system to ensure it is functioning optimally.

- Check for loose connections, damaged cables, and outdated components that may affect energy efficiency.

Step 10: Educate Household Members

- Educate family members or occupants on energy-efficient practices, such as turning off lights when not needed, optimizing thermostat settings, and using devices responsibly.

8.4 Time-saving techniques for wiring projects

Time-saving techniques for wiring projects can help streamline the installation process, reduce labor, and complete the project more efficiently. Here are some time-saving techniques to consider:

1. **Proper Planning**

- Plan the wiring project thoroughly before starting. Create a detailed layout, identify the locations of outlets, switches, and appliances, and determine the most efficient wiring routes.

- Organize all the necessary tools, materials, and equipment beforehand to avoid interruptions during the installation.

2. Use Prefabricated Wiring Components

- Consider using prefabricated wiring components, such as pre-cut cables and wire harnesses. These components save time on cutting and stripping wires on-site.

- Prefabricated components also help reduce the risk of errors and ensure consistent quality.

3. Labeling and Color Coding

- Label wires and cables clearly to avoid confusion during installation. Color-coding wires, based on their functions, can also speed up the process, especially in complex wiring systems.

- Proper labeling and color-coding help with troubleshooting and future maintenance.

4. Use Quick Connectors and Push-in Terminals

- Utilize quick connectors and push-in terminals for electrical connections where applicable. These connectors allow for faster and more secure wire connections without the need for twisting and wire nuts.

- Quick connectors can save time, especially when dealing with multiple connections in junction boxes.

5. Group Similar Tasks

- Group similar tasks together to minimize setup and cleanup time. For example, install all outlets in a room before moving on to switches or lighting fixtures.

- Avoid jumping between different tasks, as it can lead to inefficiency.

6. Power Tools and Equipment

- Use power tools like cordless drills and wire strippers to speed up repetitive tasks.

- Electric staple guns can be helpful for securing cables quickly, especially in hidden locations.

7. Cable Pulling Techniques

- Employ efficient cable-pulling techniques, such as using fish tape or cable-pulling lubricants, to minimize effort and time in running cables through walls or conduits.

8. Collaborate with a Team

- On larger projects, work with a team of qualified electricians to divide and conquer the tasks. Collaboration allows simultaneous progress on different aspects of the project.

9. Invest in Automation and Smart Devices

- Consider using automation and smart devices in the wiring system. Smart switches, outlets, and dimmers can simplify installation and save time on configuration.

10. Regularly Maintain Tools

- Keep all tools well-maintained and in good working condition. Dull blades or malfunctioning tools can slow down the project and lead to mistakes.

Chapter 9: ZigBee wireless technology and home automation

The term "home automation" refers to a system that consists of a series of electrical, communication, and device interfaces. This system uses the Internet to connect commonplace items to one another. You can control any device, no matter where you are, as long as it has detectors and a WiFi connection. This means that you can use a mobile device like a smartphone or tablet, even if you are in a whole other country.

The controlled and computerized control of processes, activities, and technology in a house is referred to as "home automation," and the phrase "home automation" was used to characterize this process. Simply said, you may control the appliances and amenities in your house remotely to enhance the convenience of your home, boost its level of safety, and even reduce the amount of money you spend on maintaining it. Keep reading to find out the responses to some of the most commonly asked concerns about technology for home automation and to obtain some recommendations for remedies to problems that might be caused by home automation.

How does one go about automating their home?

Controllers, detectors, and actuators are the three primary elements that make up a home automation system.

- Changes in temperature, sunlight exposure, or even movement may be detected via sensors. After that, your home remedies system will be able to adjust these settings as well as any others in line with your particular preferences.

- The equipment, such as computers, tablets, or cell phones that is utilized for sending messages about the operation of automated components in your home is referred to as "controllers," and the phrase "controllers" is what is meant by the term "controllers."

- Actuators are the components of a home automation system that are responsible for controlling the real workings or function of the system. Actuators might be light switches, motors, or powered valves. They are configured in such a way that they may be activated by the remote command of a controller.

9.1 What kinds of capabilities do home automation systems provide their users?

The variety of services and possibilities offered by home automation systems is rather extensive. The automation of security systems and recording devices, remote control of lights and thermostats, remote management of appliances, and detection of carbon monoxide are all examples of smart home features. Notifications are sent by SMS and email in real-time, together with live video monitoring and alarm systems. Integration with digital assistants, keyless entry, and control by voice command.

What are some of the benefits that home automation may provide?

The primary purpose of a system for home automation is to make routine domestic tasks easier to perform. Think about some of these benefits for a moment:

You can control your home using a mobile device such as a tablet, laptop, or smartphone if you have remote access.

- **Energy conservation:** Home automation gives you the ability to be more conscious of how much electricity your home uses. To save money on your energy bill, try turning down the thermostat in a room before you leave it or reducing the length of time that the lights are left on throughout the day.

- **Convenience:** You may configure devices so that they turn on automatically at specific times or so that they can be accessed remotely from any place with an Internet connection. When you don't have to worry about forgetting to lock the door behind you or turning out the lights, you'll have more mental energy to devote to more important matters.

- **Comfort:** Through the use of home automation, you may make your home cozier and friendlier to guests. You may arrange your smart speakers to begin playing music when you get home from work, program the thermostat with your preferred settings to ensure that the temperature in your home is always comfortable, or adjust the level of brightness or transparency of your lights depending on the time of day.

- **Increased safety:** Automated home security technologies like sensors for pressure, carbon dioxide monitors, and intelligent fire detectors may assist in protecting your property from calamity in the event of an emergency.

9.2 What Is the ZigBee Wireless Technology, and How Does It Work?

It is a kind of wireless technology that, like Bluetooth and Wi-Fi, permits the linking of a number of different smart home products to one another. This wireless technology is based on the EEE 802.15.4 specification as its primary building block. It had a lower power consumption and was developed for communication at close range. Even while this particular kind of wireless communication has been present for over a decade, the rise of automated homes is what is responsible for its current popularity. You may connect an intelligent light switch to a hub or system in your own home by making use of ZigBee technology, for instance. There is also the possibility of connecting a digital bulb of light to the switch. The ZigBee wireless protocol makes it easy to join the devices, even if those devices were manufactured by a variety of different companies. The fact that this protocol uses a mesh network is the primary advantage that it offers. This is perhaps the most important reason why such a large number of individuals use its usage for the automation of their homes. You may put it to use to expand not just the capabilities of the device but also the region that the network can service. According to information provided by the ZigBee Alliance, an increasing number of companies are developing intelligent devices that are compatible with the protocol. As a direct consequence of this, you can now easily link various gadgets across your home by using the same protocol. The coordinators, routers and switches, and end devices are the three distinct types of computing hardware that are used by this method. The coordinator function is performed by the primary or local network. Routers are devices that are both intelligent and fully powered, and they are used to repeat a signal. End devices are remote electronic contraptions that are powered by wireless transmission and are connected to the network by means of routers.

9.3 ZigBee is Being Used for Home Automation

ZigBee has emerged as the standard for home automation on a worldwide level. It is possible to utilize it to construct intelligent homes. It allows you to control smart equipment, which may help you conserve energy and increase the security of your house. Through the application of technology, almost any home might be converted into a smart home. These days, there is widespread interest in energy efficiency among people. Because of this, they choose appliances that have a low impact on the environment. As a result, you may reduce your energy consumption by constructing a mesh network using smart devices in conjunction with the wireless ZigBee technology. If you want to automate your house with the help of ZigBee, be sure to follow these steps.

- **Step 1: Setup ZigBee Coordinator**

The first thing that will need to be done on your end is the installation of a ZigBee coordinator. It is the most important piece of hardware in the home, and it functions as a network. You may go out and get something called a Wi-Fi mesh modem, then set it up in your home. It is possible to configure a home control hub such that it also functions as a coordinator. The ZigBee network's primary communication hub will be a part of it.

- **Step 2: Connect ZigBee Router**

After you have purchased and set up your ZigBee router, you will be able to connect the ZigBee Coordinator to it. Any intelligent device that conforms with the protocols of ZigBee and has a battery that is still functional may be used. A programmable light switch in the smart home may be linked to the coordinator. The computerized switch will now become part of the home network and will serve in the capacity of a repeater as well as a router. You are able to join many smart home appliances when you use the mesh modem in conjunction with the Zigbee network protocol. Each and every one of the devices will be linked together.

- **Step 3: Connect Other ZigBee Devices**

The network is capable of accepting more ZigBee smart devices as connections. If the electronic device is too far away to reach, the routers may still be utilized to extend the connection for a longer period of time. Building a solid mesh network for home automation that can support many devices will be a breeze with this approach. It is possible to connect to the router a smart detector, a smart security camera, a smart bulb, and a plethora of other smart devices. Through the use of ZigBee in this way, you are able to automate your home. The network will make it possible for you to connect all of the intelligent home appliances that you have. Every device in your home may be operated by your phone using either a voice control or a remote control app.

When using ZigBee, it is possible to connect more devices while still preserving a significant amount of power than would be otherwise possible.

9.4 What exactly is meant by the term wireless home automation?

Wireless home automation is the term given to the autonomous administration of internet-connected home devices and appliances. It gives you the ability to tailor the schedules of your devices, regulate when they turn on and off, and adjust numerous options to suit your preferences. A home with an intelligent management program also gives you the ability to regulate every facet of your home from a distant location. If you have a thermostat that is smart, for example, you can configure it to adjust the temperature for a period of time when you wake up in the morning, and then it can be scheduled to shift to a setting that uses less energy when you leave home for the day. Smart light bulbs may be configured to come on at a certain time, such as when you want to get up in the early hours, and they may be designed to turn off at a specific time, such as when you are ready to go to bed. Because you will be using a smaller amount of time and effort on manual work, this may make your day-to-day routines go more smoothly, allowing you to spend less money on energy while doing so. In addition, automated lighting systems provide an added layer of protection by turning on the lights inside and outside the home at predetermined intervals throughout the day or when unwelcome guests are present on the property (such as a raccoon going through your garbage or the neighbor's cat stealing goodies from the garage). It is also possible to employ intelligent safety lights to detect when someone from the delivery team delivers groceries or meals to your home. Appliances in smart homes may potentially be able to coordinate their actions with one another thanks to the network of linked IoT devices. For example, if a particular motion detector inside or outside of the house senses movement, a smart camera system may turn on the lights both inside and outside the home.

Utilizing a Wi-Fi Module for an Intelligent Home Automation System

The home automation system is able to detect moisture, temperature, and the intensity of the light using its Cayenne IoT (or Web of Things) platform. Additionally, it is able to control two electrical devices. It's possible that the two electrical objects in question are a fan on the ceiling and a light bulb, but they may just as well be any other couple of electrical devices.

Basic IoT components

Some of the essential components of an Internet of Things system include devices such as sensors and actuators, embedded

computer systems, network connectivity, user interfaces, and large-scale data storage.

- **Actuators**

Relays that are connected to output pins serve as the actuators that are employed in this scenario. Electrical loads such as fans and lights are connected to the relays' acquaintances, which allow for remote control of such loads via a web interface or a mobile application.

- **Sensors**

This project makes use of two different sensors in its implementation. One of them utilizes a light-dependent resistor, or LDR, to determine the level of ambient light intensity, while the other makes use of a DHT11 humidity and temperature sensor.

- **Embedded system**

This project makes use of an integrated controller in the form of a Wi-Fi module that also contains an ESP8266 (NodeMCU). This controller is able to process either analog or digital information that is received by sensors and then sent over the internet. Arduino is used to write the code for it. At the same time, it processes commands received from the internet and reacts to those orders by activating the actuator or other connected equipment.

- **User interface**

The Transmission Queue Receiver Transport protocol, often known as MQTT, is one that the Cayenne platform is able to implement. MQTT is a straightforward messaging system that operates over the TCP/IP connection and uses the protocol called MQTT. It is designed to work with devices that have limited bandwidth and power requirements. Using the Cayenne platform's intuitive drag-and-drop functionality, a dashboard can be rapidly created in a short amount of time. As a direct consequence of this, the time and effort required to build the user interface are reduced.

- **Network link**

The Internet is the network link that establishes a connection between the system that has been embedded and the outside world.

- **Circuit and working**

Demonstrates the schematic for a home automation system based on an ESP8266 microcontroller. Its construction is comprised of a NodeMCU (ESP8266), an LDR, a DHT11 detector, an opt mixer 4N33 (IC2 to IC3), two sets of switches (RL1 and RL2), and a few additional components. A digital input may be used to get measurements of temperature and humidity levels from the DHT11, and an analog input can be used to obtain readings of light intensity from the LDR1. To measure a broader range of temperature and humidity, first, replace the DHT11 sensor with the DHT22 sensor and then replace the DHT22 sensor with the DHT11 sensor again. The relationship of DHT11 versus DHT22 is presented in tabular format below for your convenience. The other requirements, such as the arrangement of the pins, have not been altered. Because the circuit has power relays that operate at 12V, it receives its power from a source that operates at 12V DC. A 7805 regulator takes in 12V and produces 5V so that it may power an ESP8266 NodeMCU module. The NodeMCU V1.0 (ESP8266 ESP-12) has a total of 11 GPIO pins in addition to an ADC input pin with a resolution of 10 bits. The information on the NodeMCU module's pins may be seen in Figure 4. It has a USB-to-serial conversion based on the CP2102 built in, which makes it simple to interface with a personal computer in order to upload code written in Arduino to NodeMCU. Additionally, it has an integrated 3.3V voltage regulator. On the NodeMCU, you will find a male header with 30 pins (215). Its accompanying components are linked either on a general-purpose printed circuit board (PCB) or in accordance with the architecture of the desired PCB. The dashboard of the Cayenne vehicle displays a percentage of the light intensity that was detected by the Hdr in the vehicle. This attaches to the A0 pin of the NodeMCU module and gives you the ability to read the digital voltage that is produced depending on the amount of light in the environment. The sensor known as DHT11 is used to monitor the relative humidity and temperature of the environment, and it does so by connecting with digital I/O port D5 (GPIO14). This sensor has the capability to split data for temperature and relative humidity, and the two sets of data may be communicated using the same output pin. Two BC547 transistors are in charge of controlling two 12V relays by way of MCT2E/4N33 optoisolators. Each relay has termination headers that are connected to it, and these headers may be utilized to drive an AC or DC load.

Chapter 10: Bonus Topics

Each and every important piece of information has been given in this book. Any one may learn wiring after carefully reading the complete book. This chapter will help you to advance your techniques for wiring.

10.1 Time-saving techniques for wiring projects

When it involves tasks involving wiring, efficiency is essential to attaining better outcomes and saving more time overall. In order to save time on wiring jobs, here are some time-saving techniques:

- **Plan and Prepare**

Invest some time in planning and preparation before beginning the job of actually wiring the device. Construct a comprehensive wiring plan or layout, determine the most time- and cost-effective paths for the wires, and collect all of the required equipment and components.

- **Use Wiring Conduits**

By using wire conduits or cable channels, the procedure may become more structured, which, in turn, will save time throughout the course of the project. They shield the wires and make it simpler to route them in an orderly fashion, hence lowering the likelihood that the wires would get tangled or confused.

- **Pre-cut and Label Wires**

If it's at all feasible, pre-cut the wires and identify them with the areas where they are supposed to go. During the installation process, you will be able to identify the proper location of each wire more quickly and prevent making needless cuts or changes thanks to this.

- **Use Zip Ties and Cable Clips**

Use cable clips and zip ties to secure the wires as they go along their respective pathways. This prevents them from becoming knotted up, which in turn helps keep things nice and organized. When working with longer wire lengths, cable clips might prove to be an exceptionally beneficial tool.

- **Group Wires by Function**

Organize the cables according to their purpose or their destination. The installation procedure may be streamlined, and the potential for mistakes can be reduced by grouping cables that perform the same function together.

- **Invest in Quality Tools**

It is possible for a very little investment in high-quality equipment, such as wire strippers, crimpers, and testers, to provide a considerable improvement in productivity. They make the work simpler to do and require less time overall.

- **Consider Modular Wiring Systems**

Installing modular wiring systems may sometimes be done more quickly than traditional methods. These systems make use of connectors and components that are pre-made and easily snap together, hence eliminating the need for individual wire connections.

- **Practice Proper Wire Management**

Maintain a safe distance between the wires and any moving components or possible dangers. Install them in such a way that they are attached to walls or other sturdy structures to prevent damage and provide a clean installation.

- **Automate Where Possible**

Consider using automation equipment for projects that are more sophisticated, such as wire labeling machines or wire cutters, so that you can perform repeated chores in a rapid and correct manner.

- **Double-Check Continuity and Connections**

Taking a few additional minutes here and there to double-check continuity and connections as you go may save time that would otherwise be spent troubleshooting and resolving problems later on. This may sound paradoxical, but it is true.

- **Use Wire Color Codes**

If they apply, comply with the requirements for the color coding of the wires. This will save you time by allowing you to immediately determine the purpose of each wire without requiring you to continuously go to the documentation.

- **Follow Safety Guidelines**

Taking necessary preventative measures is very necessary in order to stay out of harm's way and save money. Carry out the task in a systematic manner, and if you have any uncertainties, either speak with an expert or refer to the relevant rules. Keep in mind that while these methods might cut down on the amount of time needed, it is imperative that accuracy and safety take precedence over speed.

10.2 Avoiding common mistakes and pitfalls while wiring

To ensure a secure and efficient installation, it is essential to steer clear of typical errors and traps that might occur throughout the wiring process. The following are some pointers that can assist you in avoiding possible difficulties:

- **Read and Understand the Wiring Diagram**

Before getting started, you need to get familiar with the wiring schematic as well as the layout. Incorrectly understanding the diagram might result in broken connections and other problems with its operation.

- **Turn off Power**

Always switch off the power before working on a circuit or piece of equipment that uses electricity. Electrical shocks, as well as short circuits, may be avoided as a result of this measure.

- **Use Proper Tools and Equipment**

Ensure that you are in possession of the appropriate tools for the task at hand and that those items are in excellent functioning condition. When tools are broken or not properly maintained, it may lead to errors and accidents.

- **Avoid Overloading Circuits**

It is important to ensure that loads are distributed uniformly and do not exceed the capacity of the wire and the circuit breakers. The overloading of a system may both start fires and cause damage to the electrical infrastructure.

- **Secure Wires Properly**

When securing cables, be sure you use the right cable clips, ties, and conduits. If you leave them dangling or just loosely attached, you run the risk of causing harm or accidentally disconnecting them. Avoid doing this.

- **Properly Strip and Crimp Wires**

When stripping wires, use extreme caution so that the conductors are not harmed. Ensure you crimp connectors firmly to avoid loose connections.

- **Label Wires and Connections**

During the installation process, clearly labeling the wires and connections will save you time during subsequent maintenance and repairs. A label that is easy to read and understand helps prevent misunderstanding and makes identification simple.

- **Test Continuity and Connections**

Conduct continuity tests and double-check all connections before putting the finishing touches on the installation. This phase assists in spotting any problems early on and ensures that everything is linked in the correct manner.

- **Follow Building Codes and Standards**

Wiring should be done in accordance with both regional construction requirements and industry standards. Failure to do so may result in health and safety risks, as well as possible legal complications.

- **Keep Wiring Away from Hazards**

Wiring should not be routed near any components that release heat or move, as well as any sources of water. Take special precautions to avoid causing any physical harm to the wires.

- **Use Wire Nuts and Electrical Tape Properly**

When using wire nuts, check that they are the correct size for the wire and that they are twisted firmly. Electrical tape should be applied over any exposed wire ends to avoid short circuits.

- **Don't Rush**

During the course of the installation, ensure that you take your time and thoroughly examine your work. When you're in a hurry, you could make errors that need more work and effort to remedy later.

- **Consult Manuals and Experts**

If you come across any equipment or systems that are foreign to you, it is recommended that you check the manuals provided by the manufacturers or seek the guidance of experienced specialists.

- **Test the System After the Installation**

When everything is in place, a comprehensive test should be performed on the system or circuit to ensure that it is both functional and safe.

- **Keep a Clean and Organized Workspace**

Keep your work area clean and organized to reduce the risk of inadvertent harm and enhance your ability to concentrate on your tasks. You may considerably cut down on the possibility of making mistakes and guarantee that your wiring project will be completed successfully and without incident if you pay attention to the following examples of typical oversights and dangers.

10.3 Recommended resources for further learning

The following are the recommended sources for further learning:

1. **Online Courses**

- Udemy: Search for courses in electrical wiring that are offered by well-regarded teachers.
- Look for classes on electrical installations and other wiring procedures on the website Coursera.
- On LinkedIn Learning, you may search for video lessons and classes pertaining to the field of electrical work.

2. **YouTube Channels**

- Ask This Old House provides helpful advice and guidance that are broken down into steps for a variety of electrical repairs.

- Home Repair Tutor is an online resource that offers lessons on a variety of issues related to home maintenance, including electrical work.

- Electrical Industry Network offers video content that covers electrical product evaluations, as well as electrical theory and procedures.

3. **Websites**

- The Spruce has a section on electrical wiring as well as electrical tasks that may be done by the reader themselves.

- Electrical Safety Foundation International (ESFI) offers information and tools on how to be safe around electricity.

- Electrical101 is a website that focuses on teaching users the fundamentals and ideas of electrical wiring.

4. **Forums and Communities**

- ElectricianTalk is an online community where professional electricians and do-it-yourself electrical enthusiasts debate a wide range of electrical subjects.

- Electricians is a community on Reddit that enables professionals and hobbyists in the field of electrical work to share their expertise and experiences with one another.

5. Manufacturer's Resources

- If you are looking for product-specific information and resources, it is recommended that you look on the websites of respected electrical equipment manufacturers.

6. Local Workshops and Trade Schools

To find electrical classes and workshops in your area, you can check with local trade schools, neighborhood colleges, and vocational institutes.

7. National Electrical Code (NEC) Handbook

The National Electrical Code, which determines what constitutes a safe electrical installation in the United States, is dissected and analyzed in great detail in the NEC Handbook. Keep in mind that dealing with electricity may be dangerous, and it is very necessary to make safety your first priority. Before trying difficult electrical jobs on your own, it is a good idea to get some hands-on instruction or seek the advice of a professional electrician if you are unsure of your ability. Always ensure that you are in compliance with the local safety standards and construction requirements.

SCAN THE E-BOOK BONUS

SCAN THE CHRISTMAS

SCAN THE IMAGES

Conclusion

It is essential that correct wiring procedures be followed in order to guarantee the security, efficacy, and dependability of electrical systems. Individuals are able to establish an electrical infrastructure that is well-organized and functioning if they have a fundamental grasp of wiring, which includes knowledge of wire diameters, color codes, and suitable terminations. It is very necessary to educate oneself on proper wiring methods and procedures before beginning any kind of wiring job. Individuals who possess this expertise are able to effectively design and carry out wire installations, hence reducing the likelihood of experiencing electrical risks and assuring compliance with applicable electrical rules and regulations. In addition, the implementation of energy-efficient wiring methods may greatly cut down on the amount of power used and help the creation of a living environment that is more environmentally friendly. Homeowners may conserve energy, lessen their environmental imprint, and cut their utility costs all at the same time by installing LED lighting and smart controls and optimizing the routing of their cable systems.

It is essential to do routine maintenance and safety inspections in order to detect and address any possible wiring problems in a proactive manner. Troubleshooting and maintenance performed at the appropriate times may reduce the likelihood of accidents, increase the useful life of electrical components,

and save money on expensive repairs in the future. When executing specialty wiring projects in settings such as industrial, agricultural, or maritime environments, it is essential to comply with the norms and safety regulations particular to the relevant industry. Specialized applications need tailored solutions that are created to meet the requirements that are particular to those applications in order to guarantee that electrical systems in these settings are both reliable and efficient. In addition, the use of smart home wiring and automation technology enables increased convenience, comfort, security, and energy efficiency in the home. The integration of voice control, home automation systems, and smart devices enables seamless control of electrical components and gadgets, which in turn improves the living experience as a whole.

Individuals are given the ability to develop safe, dependable, and intelligent electrical systems when they have a complete grasp of wiring procedures, practices that are energy efficient, and specific applications. Adopting these principles will guarantee that wiring projects are carried out effectively, which will, in turn, promote sustainability, comfort, and safety in residential, commercial, and industrial environments. Individuals are able to continue reaping the many benefits of a well-planned and intelligently executed electrical infrastructure if they adhere to correct wiring standards and remain current with technology improvements.

The method in which we approach electrical systems and safety has undergone a substantial sea change as a direct result of technological advances in electrical wiring. Researchers, engineers, and manufacturers have, over the course of many years, exerted a great deal of effort toward the development of novel solutions that enhance electrical systems' levels of efficacy, sustainability, and overall safety. This unceasing advancement has fundamentally altered the manner in which we provide electricity to our homes, places of business, and industries, therefore propelling us toward a more connected and electrified future.

The creation of smart and intelligent wiring systems is one of the most important technological advancements that have occurred in the field of electrical wiring technology. These systems make use of state-of-the-art sensors, microprocessors, and communication protocols in order to establish intelligent networks inside buildings and across infrastructure. The capacity of smart wiring systems to monitor energy usage, identify defects, and optimize electrical distribution has shown to be a vital factor in boosting energy efficiency, decreasing waste, and eventually lowering the cost of utility services.

In addition, developments in materials science have allowed for the production of wire components that are both more secure and more long-lasting. In order to lessen the likelihood of fires and electrical mishaps, materials for insulation that are non-

toxic and resistant to flame have been developed. These materials guarantee that even in the event that there is a defect, the danger of fire spreading is minimal, so protecting both lives and property from harm.

Additionally, in recent years there has been a rise in the popularity of environmentally friendly and sustainable wiring solutions. Wiring items that are built from recycled materials and intended for minimal environmental effects are now available to individuals and companies who are environmentally aware and want to reduce their influence on the planet. Advanced wiring technologies make it simpler to incorporate renewable energy sources like solar power and wind turbines into electrical systems. This encourages a more sustainable and environmentally friendly approach to the production and use of electricity.

Electrical installations continue to place a premium on safety, and recent developments in wiring technology have made important contributions to the development of more foolproof safety protocols. Innovative safety devices such as ground-fault circuit interrupters (GFCIs) and arc-fault circuit interrupters (AFCIs) are examples of devices that can quickly identify and interrupt electrical faults, therefore lowering the danger of electrical fires and electric shocks. Having these devices

installed is now required by many construction regulations, which contributes to an overall improvement in the safety of electrical systems.

Another significant development that has impacted how wiring technology is used is the shrinking of various electrical components. Connectors and switches that are both smaller and more efficient have made it possible for electrical systems to be both more flexible and easier to manage, especially in areas that are both complicated and constrained. This has proved of great help in the creation of electrical gadgets that are smaller and lighter in weight, from cellphones to medical equipment, without sacrificing either their performance or their level of safety.

A new age of connection and automation has begun as a result of the emergence of the Internet of Things (IoT), which has also had an effect on the landscape of wiring. The Internet of Things enables devices to interact with one another and to be managed remotely, enabling possibilities for energy management and convenience that have never been seen before. The use of energy may be optimized, occupants' safety can be improved, and the general quality of life can be elevated in smart houses that are outfitted with Internet of Things devices.

Wire testing and diagnostics have also been significantly impacted by developments in technology. Thermal imaging and capacitance measurements are two examples of non-destructive testing procedures that enable electricians to spot flaws and possible concerns without pulling down walls or dismantling electrical panels. This speeds up the process of identifying electrical issues while also reducing the amount of interruption they cause.

The importance of electrical wiring technology is only going to grow as the globe progresses toward a more electrified and connected future in future. The continued research and development that is being done in this subject promise to produce even more fascinating ideas and solutions, which will eventually reshape the way that we harness and distribute electrical power. However, in spite of all the developments, it is vital to keep in mind the difficulties and possible dangers that are associated with sophisticated electrical systems. Maintenance on a regular basis, inspections at regular intervals, and strict adherence to established safety regulations are still absolutely necessary in order to guarantee that electrical infrastructure will continue to operate safely and effectively.

The ever-advancing technology of electrical wiring has raised the bar for the quality of electrical installations, elevating them to a higher level that is safer, more efficient, and friendlier to the environment. These developments have brought about considerable beneficial improvements in the electrical sector, and one example of this is the growth of smart wiring systems. Other examples include eco-conscious materials, enhanced safety devices, and the integration of IoT. Let us make sure that, while we welcome these technological developments and look forward to a future that will be increasingly electrified, we always put safety, sustainability, and responsible use of electrical power first. Electrical wiring technology will continue to play an important part in the process of making the globe a brighter and more connected place, provided that it is implemented thoroughly and that further developments are made continuously.